中国海洋经济评论
Ocean Economics Review of China

第一辑　2008年6月　Number 1 June 2008

第二卷
Volume 2

姜旭朝　主编

经济科学出版社
Economic Science Press

责任编辑：吕　萍　陈　静　张　辉
责任校对：杨晓莹
版式设计：代小卫
技术编辑：邱　天

图书在版编目（CIP）数据

中国海洋经济评论．第2卷．第1辑／姜旭朝主编．—北京：经济科学出版社，2008.8
ISBN 978－7－5058－7526－5

Ⅰ．中… Ⅱ．姜… Ⅲ．海洋经济学－中国－文集 Ⅳ．F74－53

中国版本图书馆CIP数据核字（2008）第135172号

中国海洋经济评论
（第二卷　第一辑）
姜旭朝　主编
经济科学出版社出版、发行　新华书店经销
社址：北京市海淀区阜成路甲28号　邮编：100142
总编室电话：88191217　发行部电话：88191540
网址：www.esp.com.cn
电子邮件：esp@esp.com.cn
汉德鼎印刷厂印刷
永胜装订厂装订
787×1092　16开　13.75印张　250000字
2008年8月第1版　2008年8月第1次印刷
印数：0001—4000册
ISBN 978－7－5058－7526－5/F·6777　定价：23.00元
（图书出现印装问题，本社负责调换）
（版权所有　翻印必究）

中国海洋经济评论 Ocean Economics Review of China

主　编（Chief Editor）
姜旭朝（Jiang Xuzhao）

学术委员会（Academic Committee）
戴桂林（中国海洋大学）　　Dai Guilin, Ocean University of China
丁四保（东北师范大学）　　Ding Sibao, Northeast Normal University
干春晖（上海财经大学）　　Gan Chunhui, Shanghai University of Finance & Economics
韩立民（中国海洋大学）　　Han Limin, Ocean University of China
权锡鉴（中国海洋大学）　　Quan Xijian, Ocean University of China
孙吉亭（山东省社科院）　　Sun Jiting, Academy of Social Science of Shandong
王　淼（中国海洋大学）　　Wang Miao, Ocean University of China
徐质斌（广东海洋大学）　　Xu Zhibin, Guangdong Ocean University
杨金森（国家海洋局）　　　Yang Jinsen, State Oceanic Administration
于　立（东北财经大学）　　Yu Li, Dongbei University of Finance & Economics
于良春（山东大学）　　　　Yu Liangchun, Shandong University
藏旭恒（山东大学）　　　　Zang Xuheng, Shandong University
张耀光（辽宁师范大学）　　Zhang Yaoguang, Liaoning Normal University
张耀辉（暨南大学）　　　　Zhang Yaohui, Jinan University
周立群（南开大学）　　　　Zhou Liqun, Nankai University

编委会（Editorial Board）
姜旭朝（Jiang Xuzhao）　　方胜民（Fang Shengmin）　　韩立民（Han Limin）
权锡鉴（Quan Xijian）　　　戴桂林（Dai Guilin）　　　　赵　昕（Zhao Xin）
薛桂芳（Xue Guifang）　　　朱意秋（Zhu Yiqiu）　　　　刘曙光（Liu Shuguang）
于谨凯（Yu Jinkai）

编辑部主任（Director）
刘曙光（Liu Shuguang）

主办单位（Issuing Units）
中国海洋大学经济学院　海洋经济研究中心
Marine Economy Research Center in College of Economy, Ocean University of China
中国海洋大学教育部人文社科重点研究基地海洋发展研究院　海洋经济研究所
Marine Economy Research Institute, KRI Institute of Marine Development, Ocean University of China

中国海洋经济评论

第二卷　第一辑　2008年6月

目　录

The Performance of Western Australian Ports
　　……………………………………… *Malcolm Tull*, *Fred Affleck*（1）
A Forecasting Model of Required Number of Wheat
　　Bulk Carriers for Africa ……… *Jingci Xie*, *Masayoshi Kubo*（34）
Managing Australian Defence Force Activities in Marine Protected Areas:
　　Using Jervis Bay as a Case Study
　　………………………… *Lingdi Zhao*, *Xiaohua Wang*, *Brian Lees*（45）
论海洋发展的基础理论研究…………………… 杨国桢　王鹏举（59）
和谐社会建设指向下的政府海洋管理转型 ………………… 徐质斌（70）
中国海洋经济理论演化研究 ………………… 姜旭朝　黄　聪（83）
国际海洋保护区研究进展：一个经济学视角 ………… 刘　康（114）
中国海洋产业可持续发展：基于主流产业经济学视角的分析
　　……………………………………………… 于谨凯　李宝星（136）
胶州湾围垦行为的博弈分析及保护对策研究 ………… 孙　丽　刘洪滨（167）
国际及区域创新体系建设：理论进展与海洋创新体系实证
　　……………………………………………… 刘曙光　朱翠玲（183）
中国海洋渔业产业化发展模式探讨 ………… 邵桂兰　李洪铉　张　希（196）

《中国海洋经济评论》[2008卷] 征稿启事 ……………………（206）

Ocean Economics Review of China
Volume 2　Number 1 June 2008　CONTENTS

The Performance of Western Australian Ports
　　　　··· *Malcolm Tull*, *Fred Affleck* （1）
A Forecasting Model of Required Number of Wheat
　　Bulk Carriers for Africa ······················· *Jingci Xie*, *Masayoshi Kubo* （34）
Managing Australian Defence Force Activities in Marine Protected Areas:
　　Using Jervis Bay as a Case Study
　　　　························· *Lingdi Zhao*, *Xiaohua Wang*, *Brian Lees* （45）
Research of the Basic Theories of Ocean
　　Development ································· *Yang Guozhen*, *Wang Pengju* （59）
Transition of Government's Role in Ocean Administration in the Collective
　　Effort to Build a Harmonious Society ···························· *Xu Zhibin* （70）
Research on the evolution of the theories of marine
　　economics in China ····························· *Jiang Xuzhao*, *Huang Cong* （83）
The Literature Review of Marine Protected Areas:
　　An Economic Perspective ···································· *Liu Kang* （114）
A Sustainable Development Research on China's Marine Industry:
　　Based on the Viewpoint of the Main Industrial Economics
　　　　··· *Yu Jinkai*, *Li Baoxing* （136）
A Game Analysis of the Reclamation Behavior and Research on
　　Protecting Strategy ································· *Sun Li*, *Liu Hongbin* （167）
National and Regional Innovation System: Theoretical Approaches and
　　Empirical Studies of Marine Innovation System
　　　　··· *Liu Shuguang*, *Zhu Cuiling* （183）
Discussion on development mode of Chinese ocean fishery
　　industrialization ··················· *Shao Guilan*, *Lee Hunhyeon*, *Zhang Xi* （196）

The Performance of Western Australian Ports[①]

Malcolm Tull, Fred Affleck[*]

[**Abstract**] The aim of this paper is to undertake an analysis of the performance of Western Australia's port authorities. The context for this research is the report released in February 2006 by Access Economics (*A scorecard of the design of economic regulation of infrastructure*) for the Australian Council for Infrastructure Development. This report was critical of the regime for economic regulation of Western Australia's ports, and by implication of the potential quality and efficiency of service delivery to their principal stakeholders. However, a reading of the Access Economics report and supporting data suggests that its analysis takes no account of the regulatory frameworks for port authorities in Western Australia (WA) contained in the *Port Authorities Act* 1999 (WA) and elsewhere, or of the actual economic and physical performance of WA port

---

① This research was funded by the Department for Planning and Infrastructure WA but the authors and not the Department are responsible for the views expressed. We would like to thank Mr Sarjit Singh, Murdoch University, for research assistance with this project.

Malcolm TULL, Associate Professor in Murdoch Business School, Murdoch University South Street, Murdoch, Western Australia 6150. Tel: +61 8 9360 2481. Fax: +61 8 9310 5004 Email: m.tull@murdoch.edu.au.

* Fred Affleck, Professor in Planning and Transport Research Centre, Curtin University of Technology, GPO Box U1987, Perth WA 6845, Australia Tel: +61 8 9266 1383 Fax: +61 8 9266 1377 Email: f.affleck@curtin.edu.au.

authorities. In the light of this apparently flawed analysis of the effectiveness of port regulation in WA, it is timely to review the performance of ports under the current governance structures, and to place the Access Economics report in a broader empirical performance-based context. While there is no regime for direct regulation of access to WA's port infrastructure, it is argued that provisions in WA's legislation governing the management of ports provide much of the focus, transparency and accountability required of an adequate regulatory framework. The current dominant Australian model of public ownership, with ports acting as strategic managers subject to statutory and governmental oversight, offers a viable alternative to complete privatisation and specialised regulatory controls. Efficient ports arguably can emerge from a variety of institutional frameworks-there is no single ownership or administrative structure that fits all circumstances.

[Key Words] Australia, ports, ownership, governance, regulation, performance

JEL Classification: L33, R48

1. Introduction

The aim of this paper is to undertake an analysis of the performance of Western Australia's port authorities. The context for this is the report released in February 2006 by Access Economics (*A scorecard of the design of economic regulation of infrastructure*) for the Australian Council for Infrastructure Development (AusCID). This report was critical of the regime for economic regulation of WA ports, and by implication of the potential quality and efficiency of service delivery to their principal stakeholders. However, a reading of the AusCID report and supporting data suggests that its analysis takes no account of the regulatory frameworks for port authorities in Western Australia (WA) contained in the *Port Authorities Act* 1999 (WA) and elsewhere, or of the actual economic and physical performance of WA port authorities.

In the light of this apparently flawed analysis of the effectiveness of port regulation in WA, it is timely to review the performance of ports under the current governance structures, and to place the AusCID report in a broader empirical performance-based context. The current research has three dimensions:

• It collects data and reports on the eight ports governed by the *Port Authorities Act* 1999 (Albany, Broome, Bunbury, Dampier, Esperance, Fremantle, Geraldton and Port Hedland).

• It uses performance indicators to assess the performance outcomes of these ports, including their financial performance and relevant physical measures of output and service to principal stakeholders. As far as practicable this examination covers the twelve years from 1993 – 94 to 2004 – 05.

• It examines the regulatory and policy context in which these ports operate, including the *Port Authorities Act* 1999 and other relevant legislation and policy directives of governments.

The paper begins by outlining the broad issues involved in port governance and reform; it then examines the existing state of regulation of Australian ports; next, it describes port governance in WA. With this background, port performance and the outcomes of port reform are considered in detail; and, finally, we present our conclusions.

2. Port Governance and Reform

Historically the majority of ports have been publicly owned, owing to the perception they are natural monopolies and that potentially public ownership can prevent abuse of their market power. However, Goss (1990a; 1990b) has shown that public ownership does not necessarily prevent such abuse-specifically that economic rents can be obtained by a number of port-based actors, either singly or in combination, including public port authorities, private stevedoring firms operating within ports, and/or port workers (via restrictive work practices and overstaffing).

The powers and responsibilities of public port authorities vary widely, from 'landlord' at one extreme to 'comprehensive' at the other (Goss, 1990b). A 'landlord' authority exercises overall control over the port and plans and promotes its development, but allows private enterprise to undertake most activities within the port. By contrast, a 'comprehensive' authority may undertake the majority of activities within the port, including employing the workforce and providing stevedoring services (loading and unloading vessels). In Australia, landlord ports, where the authority rents or leases land sites to private firms and exercises overall control, have predominated.

Since the 1980s there has been a worldwide trend for governments to reduce their direct role in public utilities such as ports. In Australia this began in the late 1980s as part of the federal government's microeconomic reform policy agenda, culminating in the intergovernmental National Competition Policy Agreement (1995) which aimed to produce a coordinated approach to competition reform, with a primary but not exclusive focus on the public sector. However, attempts to introduce major reforms to labour relations and work practices on the waterfront proved controversial, and led to a nation-wide dispute in 1998. By contrast, reform of port authorities, which focussed on commercialisation, corporatisation and ultimately privatisation, proceeded much more smoothly. [1]

The only Australian states to privatise ports were Victoria (Geelong and Portland sold in 1996) and South Australia (all sold to Flinders Ports Pty Ltd. in 2001). The empirical outcomes of privatisation have been widely documented in a range of countries and industries, but there is relatively little empirical work on ports and even less specifically aimed at evaluating the outcomes of non-privatisation models for port reform (Brooks, 2005). The UK has the largest proportion of privately-owned ports, with about 70 per cent of port capacity in private hands. Since privatisation began the UK government has effectively left the running of the industry to market forces, subject to state regulatory policy that aims to "add value rather than unnecessary cost" (House of Commons Transport Committee, 2003). Central government involvement in port planning and development has taken a backseat since the demise of the National Ports Council in 1981 (Reveley and Tull, 2002). A number of concerns have however, been raised about the performance of UK's privatised ports industry, including the lack of publicly available statistical information on port activities and performance, the adequacy of health and safety regulations, and the efficiency of planning processes. Currently, the British Government's overall distribution strategy emphases sustainability, intermodal integration, environmental protection and better regional and local planning rather than privatisation, competition and deregulation (Department for Transport, 1999). The shifts in British policy suggest that the time is

---

[1] 'Commercialisation' involves clarifying the objectives of port authorities by requiring them to operate on a more commercial basis, in particular with regard to financial outcomes and freedom from government direction; 'corporatisation' involves commercialisation plus the restructuring of port authorities as separate legal entities accountable for clear financial and other objectives, governed by Boards appointed on the basis of their expertise with clear powers and functions; 'privatisation' can take a variety of forms but generally involves the sale or long-term lease of infrastructure and or operational assets to the private sector.

ripe for a detailed evaluation of the outcomes of deregulation and port privatisation in other parts of the world including Australia.

3. Economic Regulation of Australian Ports

The AusCID Access Economics' 2006 assessment of infrastructure regulation was narrowly focused and based solely on analysis of legislation and regulatory frameworks. According to its authors:

> The focus is on **economic regulation which directly sets prices or revenue for access to, or use of, services provided by infrastructure owners**.... The focus is on scoring the enabling legislation and other guidelines underpinning each regime against good regulatory design principles. [Emphasis added] (Access Economics, 2006, pp. i – ii).

So the sole concern was with the design of regimes for economic regulation intended to implement Australia's 1995 National Competition Policy affecting the behaviour of public and private infrastructure owners. The report did not examine the myriad of other legislative means by which the behaviour of port owners may be controlled in the interest of their users and other stakeholders, including the community at large. Nor did it examine the regulation of privately-owned port-based service providers such as stevedores and towage operators, despite considerable evidence that significant market power may be enjoyed by them (see, for example, Productivity Commission, 1998). Ownership of infrastructure (public or private) is not stated to be a factor in the assessment, and the report states that "public ownership does not necessary [sic] have to go hand-in-hand with poor regulatory design" (p. 10).

The value of the report's findings is limited by its acknowledgement that its assessment of regulatory design is unsupported by any established need. In this regard, the report's authors acknowledge that:

> ... economic regulation should only be used where there is evidence of persistent structural impediments to achieving efficient use of, operation and investment in infrastructure by relying on market mechanisms alone. (p. 6)

However, the Access Economics report states that:

> ... *its scorecard does not rate the decisions, or outcomes, of each jurisdiction's regulatory regime or the industry that it regulate.* (p. i).

This is consistent with its examination of the port infrastructure sector, in which it takes no account of financial or service delivery outcomes.

So what specific factors are taken into account in the Access Economics assessments? The report states that it:

> ... *focuses on the extent to which the regime in each jurisdiction is designed in a way that is* **likely to foster good decisions and outcomes**. *Good decisions and outcomes are those which encourage efficient resource allocation by appropriately balancing the need of investors to earn a* **reasonable rate of return on capital** *and the interests of infrastructure users to obtain services at minimum feasible cost.* (p. i).

Four specific design features are listed in relation to ports:
- **Independence** from government, industry and other stakeholders.
- **Focus** on efficient resource allocation.
- **Transparency** predictability and consistency of regulatory processes.
- **Accountability** of regulatory processes.

3 Rating Port Infrastructure Regulation

In spite of its admitted narrow focus, the AusCID report rated the overall regulation of ports in Australia as 'poor' (see Table 1).

WA received an overall 'very poor' rating, largely a consequence of full public ownership of its port infrastructure in the state (lack of independence) and absence of direct regulation of port infrastructure. This takes no account of other legislative provisions available to government to achieve similar outcomes, in particular those relating to Focus, Transparency and Accountability. As indicated previously, it also takes no account of actual performance outcomes.

Table 1   AusCid's rating of the regulation of Australian ports

Jurisdiction: Criteria:	NSW	VIC	QLD	SA	WA	TAS	NT	Overall
Independent	Poor	V. Good	Poor	V. Good	Poor	Poor	Poor	Fair
Focussed	Poor	Good	Poor	Good	V. Poor	V. Poor	V. Poor	Poor
Transparent	V. Poor	V. Good	Poor	V. Good	V. Poor	V. Poor	V. Poor	Poor
Accountable	V. Poor	Good	Fair	V. Good	V. Poor	V. Poor	V. Poor	Poor
Overall Jurisdiction Rating	V. Poor	Good	Poor	V. Good	V. Poor	V. Poor	V. Poor	Poor

Source: AusCID/Access Economics report, p. 10; shading added.

The evident 'high performers' in terms of the report's assessment criteria are South Australia and Victoria, the only states to wholly or partially privatise their ports:①

> In these States the prices charged for port services are monitored by an independent regulator with the option for port users to seek relief under an access regime if commercially negotiated prices cannot be agreed upon. (p. 10).

A recent evaluation of port reform in Victoria was critical of the state's regulatory regime, arguing that it "has not delivered an overall reduction in costs to shippers" and had "created a less than level playing field between Victorian ports, favouring those in private ownership" (Victorian Department of Infrastructure (DOI), 2001, pp. 73, 80 & 84 – 85). It also concluded that the regulatory framework had reduced the capacity of publicly-owned ports both to undertake new capital expenditure and also to provide for the maintenance of existing assets. It had also encouraged expansion of non-regulated services. It found that regulation did not consider service outcome performance measures and was narrowly focused on price caps for prescribed services. In this regard, it found there was some evidence that it had disadvantaged some stakeholders not using prescribed services; for example, from 1997 to 2001 Melbourne Port Corporation increased its non-regulated land rents from 28 per cent to 35 per cent of total revenue, the highest of the four capital city ports.

---

① For background see ESCOSA (2004), 'Regulation of South Australian Ports. Information Paper'. It appears likely the regulatory policies in those states were motivated by the desire to ensure that the newly privatised ports did not exploit their monopoly powers to enjoy economic rents.

4. Port Regulation and Governance in Western Australia

The research reported here has set out to assess the policy framework in which WA's ports operate and the performance it has achieved in a broader context than that of the Access Economics report. The aim has been to assess how well the WA policy framework is achieving "efficient use of, operation and investment in infrastructure", which was acknowledged by the authors of the Access Economics report (p. 6) to be key criterion in establishing the need for formal direct regulation.

The remainder of this paper examines three principal areas:

- The full range of legislative means available to government for managing the performance of WA's publicly-owned port infrastructure and for regulating the behaviour (including pricing) and accountability of port infrastructure owners.
- The financial performance of WA port infrastructure owners, to assess whether there is evidence of excessive (or inadequate) financial performance.
- The service performance outcomes produced.

4.1 Economic Regulation

In the port context, the Access Economics assessment criteria would require a regime for economic regulation providing rules for a 'negotiate/arbitrate' system to facilitate access to the port by parties wishing to use port infrastructure, e.g. stevedores, pilots, other land users and service providers, and to regulate prices charged to these and other port users, eg wharfage.

There is no formal direct regulation of access to port infrastructure in Western Australia or pricing of the use of port infrastructure. In WA an independent body, the Economic Regulation Authority (ERA), was established in 2004 to administer industry-specific legislation regulating third party access to electricity, gas, rail and water infrastructure (currently the ERA's legislative charter does not include ports). However, the state government may request the ERA to inquire into matters relating to regulated or non-regulated industries, including pricing, quality, business practices and compliance costs.

4.2 Governance of Port Authorities

In Western Australia, governments have consistently rejected 'corporatisation' and 'privatisation' in favour of 'commercialisation' (Tull, 1997). The *Port Authorities Act* 1999 establishes and governs the business of WA's eight port authorities[①], and provides a framework within which the government may control matters such as prices, investment and appointment of governing authorities and key staff. The 1999 Act was a major advance in standardising the objectives and accounting standards of WA ports, although some elements of commercialisation had earlier been applied to Fremantle and Bunbury-in 1995 the government had spelled out the role of the Fremantle Port Authority (FPA), "to facilitate trade in an efficient and commercial manner", and in 1996 approved formal commercialisation of the FPA and the Bunbury Port Authority. The *Port Authorities Act* 1999 repealed seven pre-existing Port Authority Acts and the *Port Functions Act* 1993 to create a common approach to port authority management. Section 30 of the Act prescribes the functions of a port authority:

• To facilitate trade within and through the Port and plan for future growth and development of the port

• To undertake or arrange for activities that will encourage and facilitate the development of trade and commerce generally for the economic benefit of the State through the use of the Port and related facilities

• To control business and other activities in the Port or in connection with the operation of the Port

• To be responsible for the safe and efficient operation of the port

• To be responsible for the maintenance and preservation of vested property and other property held by it

• To protect the environment of the port and minimise the impact of port activities on that environment.

The Act provides some variations for the ports of Dampier and Port Hedland mainly to allow the major users (mining companies) to nominate Directors to the boards of the port authorities. Significantly, the Dampier Port Authority's objectives

---

① Albany, Broome, Bunbury, Dampier, Esperance, Fremantle, Geraldton and Port Hedland.

omitted any reference to making a profit:

> *The functions of the port authority include recovering as far a possible, the cost of the facilities and services provided by the port authority from the users of those facilities and services (Schedule 6, 1.9).*

According to WA's Department for Planning and Infrastructure:

> *The [1999] Act commercialised port authorities with an intent to better equip them to respond to market forces and thereby facilitate trade. The Act intended that ports be given the freedom to control the day-to-day running of the port, while allowing Government to retain strategic control, including the ability to set performance goals and broad limits for capital expenditure and to control the range of activities undertaken.* ①

In addition to the requirements of the Port Authorities Act 1999, ports must also comply with a wide range of regulations including state and national competition laws, infrastructure planning and building codes, financial audit and reporting and environmental legislation. Annual reports are tabled in Parliament and generally provide a high level of disclosure of both financial and non-financial performance.

4.3 Focus, Transparency and Accountability

A number of features of the 1999 Act provide partial responses to the assessment criteria used in the Access Economics report. There is a strong focus on efficient resource allocation, transparency and accountability.

Port authorities are required to "act in accordance with prudent commercial principles [and] endeavour to make a profit" (Section 34). Pricing guidelines in the Act are consistent with this: "Port charges are to be determined by the port authority in accordance with prudent commercial principles and may allow for ... the making of a profit [and] depreciation of assets" (Section 37). Each port is required to pre-

---

① Source: http://www.slp.wa.gov.au/statutes/swans.nsf/ PDFbyName/03FFB12DE1A1E936482567 D2002BA6F0? OpenDocument. Accessed 24/11/06.

pare and publish an annual Statement of Corporate Intent (Section 60), which must include estimates of revenue and expenditure, borrowings and dividends. There also a requirement to describe "pricing arrangements", the nature, costing and funding of 'community service obligations' and "performance targets and other measures by which performances may be judged and related to objectives".

Annual reports are also required and must fully disclose key financial accounting policies and outcomes, and report key performance measures. Notably, privatised entities, including ports in Australia, New Zealand and the UK, provide substantially lesser disclosure of financial and non-financial targets and outcomes.

All of the above statutory provisions are "likely to foster good decisions and outcomes", to borrow words from the Access Economics report. [1]This is even if their value may be compromised by powers of Ministerial approval and direction-an inevitable consequence of the limited 'commercialisation' model. The WA government has, however, restricted the autonomy of the port authorities in a number of areas including retention of a power of veto over charges and the appointment of senior staff.

In practical terms ports have moved a substantial distance in response to 'commercialisation', both in terms of changing business models and in their financial and non-financial outcomes. After 1996, the FPA contracted out many services including pilotage, maintenance of stevedoring equipment, and forklift driver training. Outsourcing became the accepted practice at all WA ports. Restructuring and contracting out caused the FPA's staff numbers to fall from over 750 in 1990 to less than 200 ten years later, but by the early 2000s staff numbers at Fremantle were creeping up again. Significantly, at Albany, in response to the requirements of the 1999 Act, "the Board formed the view late in 2003, that a return to the direct employment of staff would give greater control over productivity, safety and training, at the same time, offering career paths for greater job satisfaction". Albany now directly employs its maintenance and general operations staff. [2]

4.4 Competitive Neutrality

A key feature of the national competition policy reforms is a requirement for

---

[1] See the quotation at page 4.
[2] Albany Port Authority, *Annual Report*, 2003–04, p. 3

'competitive neutrality', that is, a situation in which government enterprises face the same market conditions with regard to competition, taxation and the like as competing organisations in the private sector. Under the current arrangements, all ports are required to pay dividends and income tax-equivalent payments, as well as payments in lieu of local government rates (Section 82), 'government efficiency dividends' which are not profit related. Conversely, if ports are required to undertake non-commercial activities they should receive payments in recognition of these 'community service obligations' (CSOs) to cover their net cost; Bunbury, the first WA port to receive CSO funding, was paid $85,000 in 2004 – 05 in return for providing leased areas to community-based organisations.

5. WA Ports' Performance Outcomes

A primary focus of this research is to measure the actual performance of WA port authorities-a measure of their success in meeting the expectations of their stakeholders and balancing their competing interests. The areas examined in the remainder of this paper are:

- Financial performance indicators: Return on assets; profit margin per cargo tonne; dividend payout ratio; current ratio; gearing (debt/equity ratio); and operating profit (before tax).
- Pricing: Revenue per unit of cargo; cost (i.e. expenditure) per unit of cargo.
- Non-financial performance indicators: Berth occupancy; turnround times.

To assess port performance, we have used 'performance indicators', that is measures of actual performance compared with pre-set goals related to their outputs and/or outcomes (Kearney, 1991). (These are not 'benchmarks' comparing the performance of WA ports against that of ports elsewhere.) As various port stakeholders (e.g. port users, employees and government) have differing interests, it is necessary to examine a range of performance indicators covering prices, service quality, profitability, community service obligations and employment. This methodology has previously been employed by Tull and Reveley (2001) to evaluate the performance of selected Australian and New Zealand ports.

Data sources employed include *Waterline* (Australian Bureau of Transport and Regional Economics), publications on the financial performance of government trading enterprises (Productivity Commission), and port authority annual reports.

Table 2 provides a long-term summary of key financial and non-financial indicators. The Appendix at the rear of this report contains detailed data from the eight ports governed by the 1999 Act: Albany, Broome, Bunbury, Dampier, Esperance, Fremantle, Geraldton and Port Hedland.

While the *Port Authorities Act* 1999 was a major advance in standardising the objectives and accounting procedures between WA ports, comparisons of port performance (between ports and from year to year) are nevertheless subject to many qualifications. In particular, as many of the indicators are ratios with activity-based denominators, variations in trade and shipping activity and in the scope of administrative responsibilities need to be taken into account when comparing performance. Small ports such as Albany (which handles only 3 million tonnes of cargo per annum) are unable to reap the economies of scale created by large ports like Port Hedland (which handles over 100 million tonnes of cargo per annum). Fremantle is the only port which handles high-value container traffic, while all the others are predominantly bulk ports.

Table 2   Summary Performance indicators, WA Ports
(avg. 1993 – 1994 to 2004 – 2005)

Port	Return on assets (% p. a.)	Profit margin per cargo tonne ($1989/90)	Turnaround times (Avg hours)	Turnaround times (Coefficient of variation %)
Albany	7.0	0.4	85	28.6
Broome	*	*	NA	NA
Bunbury	7.1	0.3	39	8.3
Dampier	1.8	*	24#	8.6
Esperance	9.9	0.5	46	12.6
Fremantle	14.7	0.5	26	11.9
Geraldton	9.9	0.4	50	22.7
Port Hedland	–24.1	0.1	NA	NA

Notes: * = less than 0.1; # = 3 years data only; NA = not available; Coefficient of variation = standard deviation/mean x 100.

Source: Appendix A.

5.1 Financial Indicators

From 2000 – 2001 to 2004 – 2005, the Australian port sector as a whole earned a rate of return on assets of about 6% per annum. ①Table 2 shows that from 1993 – 1994 to 2004 – 2005, *return on assets and profit margins* varied considerably between Western Australian ports. The maximum return on assets of 15% does not suggest the existence of monopoly rents, although the majority of ports are complying with the statutory requirement that they "endeavour to make a profit". In addition, the government requires ports to achieve a rate of return on assets of at least 5% – 8% per annum; only three ports failed to achieve this. ②Fremantle consistently reports the highest annual rate of return, averaging a commercial 15%, and the highest reported operating profit of all WA ports. This reflects its status as WA's major mixed cargo port and the only one with container handling facilities. Fremantle is followed by Esperance and Geraldton which earned returns averaging about 10%, and Bunbury and Albany which averaged about 7%. Broome is the worst performer but is handicapped by small cargo volumes (about 0.2 million tonnes per annum) and high fixed costs from the long jetty needed to cope with the large tidal range. In 2005 it began a $16.8 million jetty extension to allow berthing of larger vessels up to 50,000 dwt.

In Geraldton, the rate of return on assets dropped from an average of about 14% p.a. in the 1990s to about 4% p.a. after 2000; financial performance after 2002 – 2003 was adversely affected by the $103 million port enhancement project, which led to a large jump in the debt/equity ratio. Dampier's rate of return and profit margin declined after 2000, in spite of impressive 58% growth in cargo tonnages from 60.5 mtpa in 1993 – 1994 to 1995.8 mtpa in 2004 – 2005, although another key financial indicator, the current ratio (the ratio of current assets to current liabilities) improved during this time. By contrast, Bunbury's financial performance appears to have improved since 2000 with a declining debt/equity ratio and improving current ratio. Port Hedland's performance was distorted by a $134 million deficit in 2000 – 2001 caused by the write-off of channel and dredging costs; if this year is excluded the rate of return is still low but positive at 3 per cent per annum.

Turning to *profit margin per tonne of cargo* (1989 – 1990 prices), Fremantle and Esperance both averaged $0.50 per tonne, closely followed by Albany and Geraldton at $0.40 per tonne and Bunbury at $0.30 per tonne. Since 2000 Bunbury, Esperance, Fremantle and Port Hedland have maintained their profit margins in real terms.

---

① Productivity Commission, *Financial performance of government trading enterprises*, 2006.
② Department for Planning and Infrastructure, 2006.

The rate of return and profit margin data do not suggest monopoly profits are being extracted from Western Australian port operations.

As indicated above, an adverse effect of privatisation in Australia and other countries has been to reduce transparency, i. e. the amount of information available on port performance. South Australian ports, for example, are no longer monitored by the Productivity Commission.

But in 2004 – 2005, South Australia's sole port operator privately-owned Flinders Ports reported a rate of return on assets of 12.0%, an operating profit of about $16 million, profit per tonne of cargo of $0.92 and a dividend payout ratio of 97.5 per cent.① By comparison, in the same year WA's most profitable port, Fremantle, earned a slightly inferior rate of return on assets (10.5%), a similar operating profit ($16.5 million) and a lesser profit per tonne of cargo ($0.65); its dividend payout ratio was a much lower 40.1%, suggesting that more cash was left in the business to assist with investment in new capital for upgrading of facilities and for future growth.② Flinders Ports' stronger profit performance suggests there may be a link between privatisation and improved financial outcomes, but the difference does not appear to be significant based on these figures. In any case, as New Zealand's experience suggests, good financial performance benefiting shareholders may conflict with passing efficiency gains to port users (Tull and Reveley, 2001).

One of the goals of Australia's port reforms was to decrease costs and charges to port users and most Australian ports have expressed a commitment to reduce prices. In order to facilitate a comparative assessment of *trends in cost and charges* to port users, Figures 1 – 8 in Appendix B show expenditure (including debt servicing costs) per unit of cargo ($/tonne), and income per unit of cargo ($/tonne) in constant 1989 – 1990 prices for the six principal ports between 1993 – 1994 and 2004 – 2005. If performance were trending favourably, with port charges to customers decreasing and efficiency increasing, one would expect a decline in both revenues per unit of output and costs per unit of output. It is important to note that the revenue and cost data relate to *port authority* services only; aggregate data on the revenues and costs of all port service providers are not available. Due to the different characteristics of each

---

① Calculated from Australian Securities & Investments Commission, Flinders Ports Pty Ltd, *Financial Report*, 2004 – 2005. Unfortunately, the data are not sufficiently disaggregated to enable an assessment of the performance of individual ports.

② This is consistent with recommendations that port authorities be permitted to use a portion of their cash reserves for capital investment (Department for Planning and Infrastructure, 2006).

port, it may be more instructive to compare *trends* over time rather than *absolute* levels of revenues and costs per unit of output.

Table 3 below summarises the changes between 1993 – 1994 and 2004 – 2005, and shows that all ports except Dampier and Fremantle experienced falls (ranging from 15 to 30 per cent) in real costs per unit of cargo. Furthermore, comparing 1993 – 1994 and 2004 – 2005, all ports except Fremantle experienced substantial falls (ranging from 23 to 40 per cent) in real revenue per unit of cargo. Trends in unit costs and revenue at each port are shown in Figures 1 to 8 in Appendix B at the rear of this paper.

Table 3   Changes in port authority cost and revenue comparing 1993 – 1994 and 2004 – 2005 (1989/1990 prices)

Port	Change in real cost per unit of cargo (%)	Change in real revenue per unit of cargo (%)
Albany	– 25	– 23
Broome	NA	NA
Bunbury	– 21	– 28
Dampier	0	– 40
Esperance	– 30	– 33
Fremantle	– 2	– 1
Geraldton	– 28	– 39
Port Hedland	– 15	– 33

Notes: NA = not available.
Source: Appendix A.

This evidence suggests that except at Fremantle, there were significant gains in operating efficiency over this period and most importantly, it appears the majority of the gains have been transferred to port users. This seems to reflect a charging policy which does not seek to exploit market power. In this regard, the Dampier Port Authority has stated:

> ... the Authority's financial goals are secondary to its role as a trade facilitator. To that end, the aim is to minimise revenue without affecting financial viability so as to provide the most cost-effective service to port users. (Dampier Port Authority, *Annual Report*, 2003 – 2004, p. 14).

At Fremantle, comparing 1993 – 1994 and 2000 – 2001, cost and revenue per tonne of cargo fell by 29 per cent and 20 per cent respectively in real terms, but since then the declining trend has reversed. It is possible that the momentum of reform, which began earlier at Fremantle than other WA ports, has slowed. However, the FPA's corporate scorecard, based on a different methodology, claims that charges have fallen in real terms by about 40 per cent from 1993 – 1994 to 2004 – 2005 (FPA, *Annual Report*, 2004 – 2005, p. 87)[①].

5.2 Non-financial Indicators

It is widely recognised that ports compete on non-price as well as price characteristics and aspects of service quality such as speed (turnround time) and reliability can be decisive in port choice. The timeliness and reliability of port services can be gauged by examining indicators such as turnround times and berth occupancy. *Ship turnround time* captures the performance of a number of service providers including the port authority itself, pilots, tugs, stevedores and the labour force (SCNPMGTE, 1998, p. 264). Unfortunately, data on turnround times are limited, especially for non-container ports not covered by data in *Waterline*.

Table 2 above reveals significant differences between ports in turnround times. However, differences and variations in shipping and cargo volumes and composition limit the usefulness of a comparison based on absolute values. So it is more useful to examine the degree of variation and trends in this indicator of performance. The 'coefficient of variation', which measures relative rather than absolute variation, provides a better although still crude indication of reliability. Data shown in the far right column of Table 2 shows that turnround times are subject to more variation in Albany and Geraldton than in other ports. The trend in average turnround times between 1993 – 1994 and 2004 – 2005 (shown in Appendix A) appears to have improved at Esperance and Geraldton, been static at Bunbury and Fremantle (container only) and slightly deteriorated at Albany. Fremantle's turnround times improved in the late 1990s but crept up after 2003. Fremantle's turnaround times for containers are not unreasonable compared with other ports in Australia and it is important to note that all

---

① The FPA defines its real price index as "the weighted average price index deflated by the CPI for Perth. The average price equals the total of prices for individual Fremantle Ports' services weighted by their contribution to total revenue, excluding bulk cargo handling charges negotiated under commercial agreements."

container terminals are operated by private enterprise. Again one interpretation is that nationally port reforms were beginning to lose their impetus.

Data collected by the Australian Wheat Board Ltd on grain make it possible to standardise turnround times for differences between ports and for differences in average grain cargoes. Figure 1 illustrates average turnround time per 1 000 tonnes of grain lifted between 1991 – 2000. The standardised turnround times have fluctuated from year to year with Fremantle performing close to the national average, Albany and Esperance usually performing above the national average and Geraldton below. There is no clearly discernable trend although Geraldton appears to have improved its performance in the late 1990s. Analysis of the performance of the privatised Victorian ports of Geelong and Portland shows that they performed close to the national average and "it is difficult to attribute any causation from the Victorian port reforms other than to note the variability of the Victorian ports' performance has been reduced compared to the pre-reform period" (Victorian Department of Infrastructure 2001, Appendix A). These data suggest that ownership of ports may not be a critical determinant of performance.

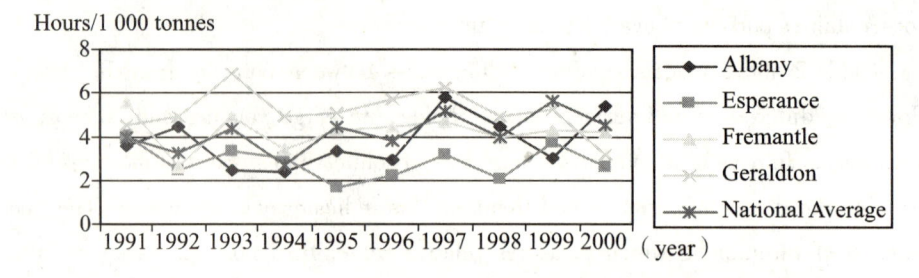

Figure 1   Average turnround time per 1 000 tonnes of grain lifted, 1991 – 2000

Source: Victorian Department of Infrastructure (2001), *The next wave of Port Reform in Victoria*, Appendix A, Meyrick and Associates Pty Ltd, Technical Paper: *Economic Impact of Port Reform*, Tables 5.11 and 5.12.

6. Conclusions

The purpose of this research has been to provide a wider perspective for assessing the appropriateness of conclusions regarding WA ports reached by the Access Economics (2006) 'scorecard' on economic regulation of infrastructure. The perspective provided here places emphasis on other regulatory arrangements enacted in legislation governing the management of WA's port authorities and on the actual performance of these port authorities. In general, the conclusions of this research are that the "very poor" scorecard rating given to WA ports is inappropriate. While the scorecard may reflect accurately the absence of a regime for direct regulation of access to port infrastructure in WA, it ignores both the more substantive issue of actual performance and provisions in WA's legislation governing the management of ports which provide much of the focus, transparency and accountability used by Access Economics in its 2006 assessment. This review of port performance does not claim to be comprehensive and gives only a snapshot of a complex situation. With this reservation, the research reported here demonstrates that:

- There is no evidence that WA port authorities are earning monopoly rents from their ownership and operation of WA's principal ports.
- There is no evidence that the charges being made for services provided directly by port authorities are excessive compared with other similar ports. Indeed, WA ports seem to take pride in holding charges as low as possible and rarely adjust them.
- There is evidence that a significant part of the financial benefits from reforms being made by major WA port authorities are being passed on to port users. The financial and non-financial indicators suggest that at Fremantle port performance peaked in the early 2000s and that regional ports are continuing to improve their performance. Performance indicators discussed in section 5 of this report show there have been significance gains in operating efficiency from 1993 – 1994 to 2004 – 2005 and that most importantly these gains have been passed onto port users.
- There is evidence that major WA port authorities are providing levels of service which are reasonable compared with other ports in Australia. While there is no regime for direct regulation of access to WA's port infrastructure, there are two alternative statutory means by which some of the objectives of such a regime can be achieved: first, via the Port Authorities Act 1999, which provides for substantial focus on economic efficiency through commercialisation, transparency and accountability, albeit potentially limited by powers of government direction; second, via the ability of government to request the WA Economic Regulation Authority (ERA) to examine the pricing, quality, business practices and compliance costs of non-regulated industries including the port industry.

There are major differences between WA ports in the scale and composition of trade and shipping flows and this complicates any comparisons of performance. The last comprehensive review of WA ports was undertaken jointly by the Bureau of Transport Economics and the WA Director General of Transport in 1981 (BTE, 1981); it is noted that a review of the WA Port Authorities Act is currently underway.

It may be noted that in relation to ports Access Economics gives the highest scores to States in which a moderate to high degree of port privatisation has occurred (Victoria and South Australia), in which arguably there is a greater need for independent regulation, owing to the absence of more direct government supervision which accompanies direct ownership.

It remains an open question whether direct economic regulation would act as a spur to increase efficiency and performance in WA's ports. The Victorian regulatory regime which received an overall 'good' rating from Access Economics' is narrowly focused on pricing and arguably has held back investment in publicly owned ports. Increasingly, it is recognised that boosting competition and efficiency does not necessarily require a change of ownership. Reform does, however, need to be a continuous process if WA's ports are to remain competitive in a globalised world economy.

References

1. Access Economics Pty Ltd, (2006) *A scorecard of the design of economic regulation of infrastructure*. Report for the Australian Council for Infrastructure Development, Canberra.

2. Breunig, R., Stacey, S., Hornby, J., and Menezes, F., (2006) 'Price Regulation in Australia: How Consistent Has It Been?' *The Economic Record*, vol. 82, No. 256, March, 2006, 67 – 76.

3. Brooks, M. R., (2005), 'Good governance and ports as tools of economic development: are they compatible?' In Lee, Tae-Woo and Cullinane, K. (eds.) (2005) World shipping and port development. Houndsmills, Basingstoke, Palgrave Macmillan.

4. Bureau of Transport Economics and Director General of Transport WA (1981), *A study of Western Australian Ports*. Canberra, AGPS.

5. Department for Planning and Infrastructure WA (2006), *Funding of Western Australian Port Authority Development Projects and State Net Debt Policy*, Maritime Policy, Transport Industry Policy Division, November.

6. Department for Transport (2000), *Modern ports: a UK policy*. London, The Stationery Office.

7. Essential Services Commission of South Australia (2004), 'Regulation of South Australian Ports. Information Paper', November.

8. Flinders Ports Pty Ltd *Copy of financial statements and reports* 2005 – 2006 (ASIC Document No. 023387069 lodged 31/10/06).

9. Goss, R. (1990a), Economic Policies and Seaports: 1. 'The Economic Functions of Seaports', *Maritime Policy and Management*, vol. 17, no. 3, pp. 207 – 219.

10. Goss, R. (1990b), 'Economic Policies and Seaports: 2. The Diversity of Port Policies', *Maritime Policy and Management*, vol. 17, no. 3, pp. 221 – 234.

11. House of Commons Transport Committee (2003), *Ports*. London, The Stationery Office.

12. Kearney, C. (1991), in J. Niland, W. Brown, and B. Hughes, *Breaking New Ground: Enterprise Bargaining and Agency Agreements for the Australian Public Service*. A Report Prepared for the Australian Minister for Industrial Relations, Canberra, AGPS.

13. 'Productivity Commission To Promote Regulatory Reform By Developing Cross-Jurisdictional Benchmarking Indicators'. Press Release, (11 August, 2006) Melbourne. Source: http://www.treasurer.gov.au/tsr/content/pressreleases/2006/086.asp. Accessed 11/12/06.

14. Productivity Commission (2006), *Financial performance of government trading enterprises*, Canberra. Source: http://www.pc.gov.au/research/crp/gte0405/gte0405.pdf. Accessed 21/1/07.

15. Productivity Commission (2006), *Performance Benchmarking of Australian Business Regulation*, PC Issues Paper Canberra. Source: http://www.pc.gov.au/study/regulation benchmarking/issuespaper/regulationbenchmarking.pdf. Accessed 23/11/06.

16. Productivity Commission (1998), *International benchmarking of the Australian waterfront*, Melbourne.

17. Regulation Taskforce (2006), *Rethinking Regulation: Report of the Taskforce on Reducing Regulatory Burdens on Business*, Report to the Prime Minister and the Treasurer, Canberra.

18. Reveley, J and Tull, M. (2002) 'Centralised Port Planning: An Evaluation

of the British and New Zealand Experience', pp. 141 – 161. In G. Boyce and R. Gorski (eds), *Resources and Infrastructures in the Maritime Economy*, 1500 – 2000. St. John's, International Maritime Economic History Association.

19. Steering Committee on National Performance Monitoring of Government Trading Enterprises (SCNPMGTE) (1998), *Government trading enterprises performance indicators* 1991 – 921 *to* 1996 – 97, *Volume* 1, AGPS, Canberra.

20. Tull, M. and Reveley, J. (2001), 'The merits of public versus private ownership: A comparative study of Australian and New Zealand seaports', *Economic Papers*, 20 (3): 75 – 99.

21. Tull, M. (1997), *A Community Enterprise: The History of the Port of Fremantle*, 1897 *to* 1997. St. John's, International Maritime Economic History Association.

22. Tull, M. (1997), 'The Fremantle Port Authority: A Case Study in Microeconomic Reform', *Economic Papers*, vol. 16, no. 4, pp. 33 – 53.

23. Victorian Department of Infrastructure (2001) *The next wave of Port Reform in Victoria-An Independent Report to the Minister for Ports*. Source: http://www. doi. vic. gov. au/DOI/Internet/Freight. nsf/AllDocs/682949DA500268B3CA256FFE001468EA? OpenDocument#next. Accessed 8/12/06.

Appendix A

Notes for all tables: Definitions: return on assets = earnings before interest and tax and after abnormals (EBIT)/average total assets; dividend payout ratio = dividends paid or provided for/operating profit after tax; current ratio = current assets/ current liabilities; Debt/equity ratio = debt/total equity.

Sources for all tables: Steering Committee on National Performance of Government Trading Enterprises, (SCNPGTE) *Government trading enterprises performance indicators* 1990 – 1991 *to* 1994 – 1995, 2 vols. (Canberra, 1996), BTCE Waterline, various issues, Productivity Commission *Financial performance of government trading enterprises*, various issues, Annual Reports of Albany, Broome, Bunbury, Dampier, Esperance, Fremantle, Geraldton and Port Hedland port authorities.

Table A.1　Selected performance indicators, Albany Port Authority

Financial indicators	Units	1993–1994	1994–1995	1995–1996	1996–1997	1997–1998	1998–1999	1999–2000	2000–2001	2001–2002	2002–2003	2003–2004	2004–2005
Return on assets	%	16.2	6.2	9.9	10.8	3.3	9.2	8.5	−1.7	0.8	3.0	7.9	9.7
Dividend payout ratio	%	0.0	0.0	0.0	35.7	0.0	38.5	28.6	0.0	0.0	0.0	0.0	0.0
Current ratio	%	14.9	10.2	2.5	1.8	4.0	3.7	4.5	1.3	0.4	0.3	0.5	0.4
Debt/equity	%	0.3	0.8	14.8	10.5	10.5	7.1	6.5	33.9	59.1	68.7	54.5	41.2
Operating profit (before tax)	$million	1.6	1.3	1.5	2.3	0.6	2.2	2.3	−0.6	−0.1	−1.5	2.3	3.0
Port authority costs/unit of cargo	$/Tonne	1.61	2.02	2.07	1.8	2.37	1.79	2.04	3.32	2.71	3.34	1.62	1.61
Port authority revenue/unit of cargo	$/Tonne	2.53	2.81	2.83	2.82	2.66	2.67	2.96	2.96	2.66	2.55	2.46	2.6
Non-financial indicators													
Total cargo throughput	million tonnes	1.7	1.6	1.8	2.3	1.9	2.6	2.5	1.7	1.6	2	2.8	3
Containerised cargo	'000 TEUs												
Average turnround time	Hours	61	60	85	124	63	77	75	84	73	80	137	100
Average total employment		24	26	26	26	26		25	5				

Table A.2 Selected performance indicators, Broome Port Authority

Financial indicators	Units	1993–1994	1994–1995	1995–1996	1996–1997	1997–1998	1998–1999	1999–2000	2000–2001	2001–2002	2002–2003	2003–2004	2004–2005
Return on assets	%								0.2		-0.7		-1
Dividend payout ratio	%								0	0.0	0.0	0.0	0.0
Current ratio	%								-	3.3	1.2	1.2	2.5
Debt/equity	%								-	0	0	-0.2	57.5
Operating profit (before tax)	$million								0.01	-0.2	-0.2		-0.4
Port authority costs/unit of cargo	$/Tonne								17.90	17.80	18.90		
Port authority revenue/unit of cargo	$/Tonne								18.00	17.00	17.50		
Non-financial indicators													
Total cargo throughput	million tonnes								0.2	0.2	0.2		
Containerised cargo	'000 TEUs								0.8	0.5	0.4		
Average turnround time (container ships)	Hours												
Average total employment											17		

Table A.3 Selected performance indicators, Bunbury Port Authority

Financial indicators	Units	1993–1994	1994–1995	1995–1996	1996–1997	1997–1998	1998–1999	1999–2000	2000–2001	2001–2002	2002–2003	2003–2004	2004–2005
Return on assets	%	11.9	12.2	8.3	5.2	6.1	4.1	5.9	7.4	5.7	6.6	5.7	6.0
Dividend payout ratio	%	3.1	2.8	0.0	0.0	11.5	10.2	35.5	30.0	50.0	50.0	46.1	55.4
Current ratio	%	386.0	1045.0	319.0	197.0	359.4	239.1	204.1	391.4	354.4	428.2	547.5	423.3
Debt/equity	%	4.7	16.1	34.3	34.0	32.0	24.5	24.1	21.8	20.2	18.2	17.9	16.3
Operating profit (before tax)	$million	4.2	5.2	3.5	2.1	2.9	1.9	4.4	6.1	4.8	5.6	4.8	5.3
Port authority costs/unit of cargo	$/Tonne	1.00	1.02	1.04	1.22	1.25	1.34	1.40	0.81	0.82	0.95	0.94	1.06
Port authority revenue/unit of cargo	$/Tonne	1.56	1.67	1.46	1.47	1.57	1.56	0.97	1.37	1.22	1.36	1.34	1.50
Non-financial indicators													
Total cargo throughput	million tonnes	7.5	7.9	8.5	8.6	8.9	9.0	10.0	11.3	11.4	11.8	11.9	12.0
Containerised cargo	'000 TEUs				45								
Average turnaround time	Hours	34	37	40	43	36	42	40	36	35	36	40	43
Average total employment		41	44	46				13	13	13	13	14	14

Table A. 4   Selected performance indicators, Dampier Port Authority

Financial indicators	Units	1993-1994	1994-1995	1995-1996	1996-1997	1997-1998	1998-1999	1999-2000	2000-2001	2001-2002	2002-2003	2003-2004	2004-2005
Return on assets	%	4.5	0.4	2.1	4.2	2.8	1.6	2.5	3.0	-1.2	1.6	1.5	-1.8
Dividend payout ratio	%	0.0	0.0	0.0	0.0	0.0	0.0	0.0	0.0	-48.5	59.8	47.3	-3.3
Current ratio	%	1.2	2.2	2.8	3.1	3.5	3.5	5.0	6.7	692.2	368.5	29	41.6
Debt/ equity	%	5.7	30.1	24.6	18.7	14.9	0.0	7.41	0.0	0.0	0.0	71.5	235.9
Operating profit (before tax)	$million	1.0	0.1	0.5	1.0	0.7	0.4	0.5	0.5	-0.3	0.4	0.5	-1.0
Port authority costs/unit of cargo	$/Tonne	0.04	0.05	0.05	0.05	0.05	0.04	0.04	0.04	0.04	0.05	0.05	0.06
Port authority revenue/unit of cargo	$/Tonne	0.06	0.06	0.06	0.06	0.06	0.05	0.04	0.04	0.04	0.05	0.06	0.05
Non-financial indicators													
Total cargo throughput	million tonnes	60.5	66.6	67.2	72.2	75.7	71.3	82.6	81.4	82.7	92.2	88.9	95.8
Containerised cargo	'000 TEUs												
Average turnround time	Hours	22	26	25									
Average total employment		11	12	11	11	10	10					14	17

Table A.5 Selected performance indicators, Esperance Port Authority

Financial indicators	Units	1993–1994	1994–1995	1995–1996	1996–1997	1997–1998	1998–1999	1999–2000	2000–2001	2001–2002	2002–2003	2003–2004	2004–2005
Return on assets	%	12.3	11.4	16.3	12.3	14.4	6.6	7.2	6.6	8.6	6.2	8.0	8.8
Dividend payout ratio	%	0.0	0.0	0.0	11.8	125.5	21.1	20.0	0.0	0.0	76.9	52.0	61.3
Current ratio	%	133.0	140.0	264.0	122.0	253.0	245.0	159.0	42.0	75.0	100.0	176.6	213.2
Debt/equity	%	99.5	93.8	115.3	92.3	77.2	56.9	50.7	146.8	236.7	224.4	206.5	187.1
Operating profit (before tax)	$million	1.1	1.1	1.9	2.3	3.2	2.9	3	2.1	3.1	1.8	3.5	4.4
Port authority costs/unit of cargo	$/Tonne	3.46	3.37	3.01	3	2.79	2.85	2.84	3.03	2.92	3.55	3.14	2.98
Port authority revenue/unit of cargo	$/Tonne	4.31	3.9	3.64	3.71	3.81	3.81	3.66	3.51	3.44	3.85	3.75	3.55
Non-financial indicators													
Total cargo throughput	million tonnes	1.3	2.1	2.8	3.1	3.1	3.1	3.5	4.3	6.2	6	7.3	7.8
Containerised cargo	'000 TEUs												
Average turnround time (container ships)	Hours	54	45	42	51	47	57	48	40	42	38	42	44
Average total employment		33	33	32	33	33	33				18*	19*	20

Table A. 6   Selected performance indicators, Fremantle Port Authority

Financial indicators	Units	1993–1994	1994–1995	1995–1996	1996–1997	1997–1998	1998–1999	1999–2000	2000–2001	2001–2002	2002–2003	2003–2004	2004–2005
Return on assets	%	14.3	15.7	14.6	14.9	20.0	17.1	15.6	13.5	14.9	13.0	11.9	10.5
Dividend payout ratio	%	0.0	0.0	0.0	0.0	10.0	10.0	19.9	20.0	48.7	41.5	50.0	40.1
Current ratio		92.8	78.1	99.6	112.4	121.3	119.6	105.5	150.0	160.4	107.9	118.9	135.5
Debt/equity	%	−248.3	−319.5	1490.2	109.1	64.9	38.8	22.9	26.8	32.6	25.9	24.5	42.5
Operating profit (before tax)	$million	8.4	9.8	8.3	10.6	17.9	13.8	14.8	15.6	19.5	17.8	17.6	16.5
Port authority costs/unit of cargo	$/Tonne	2.02	2.00	1.94	2.24	1.86	1.76	1.74	1.71	1.93	2.49	2.29	2.65
Port authority revenue/unit of cargo	$/Tonne	2.49	2.46	2.52	2.41	2.44	2.34	2.37	2.40	2.78	3.25	2.96	3.30
Non-financial indicators													
Total cargo throughput	million tonnes	20.4	20.1	21.9	18.3	21.8	23.5	23.4	22.5	22.7	23.5	25.9	25.5
Containerised cargo	'000 TEUs	169.17	189.27	198.27	209.56	250.8	275.7	300.1	354.2	383.1	431.7	466.0	468
Average turnround time (container ships)	Hours	27	30	30.7	24.8	24	23	24	22	21.5	25	28.5	27.5
Average total employment		300	226	211	196	188	175	168	167	180	205	222	236

Table A.7 Selected performance indicators, Geraldton Port Authority

Financial indicators	Units	1993-1994	1994-1995	1995-1996	1996-1997	1997-1998	1998-1999	1999-2000	2000-2001	2001-2002	2002-2003	2003-2004	2004-2005
Return on assets	%	20.3	15.2	19.1	13.6	15.5	5.4	10.5	2.8	2.1	0.9	6.0	6.8
Dividend payout ratio	%	0.0	0.0	0.0	9.5	16.7	105.4	0.0	1 800.0	0.0	117.8	3.5	0.0
Current ratio	%	363.0	267.0	269.0	213.0	252.0	488.0	266.0	275.0	131.2	76.0	128.7	219.5
Debt/equity	%	101.8	87.1	120.0	95.0	62.0	51.0	51.4	48.9	46.2	441.2	504.1	483.8
Operating profit (before tax)	$/million	2.6	1.7	2.5	2.7	1.9	0.8	2.8	0.2	0.0	0.1	4.0	2.8
Port authority costs/unit of cargo	$/Tonne	4.30	4.79	4.28	4.21	3.70	3.08	2.41	3.57	3.83	3.96	3.41	4.22
Port authority revenue/unit of cargo	$/Tonne	5.17	5.38	4.97	5.00	4.22	3.29	3.12	3.64	3.85	4.00	4.32	4.73
Non-financial indicators													
Total cargo throughput	million tonnes	2.9	2.9	3.6	3.4	3.7	3.8	3.9	2.8	2.6	2.5	4.4	5.5
Containerised cargo	'000 TEUs	77	48	65	57	49	49	41	46	38	44		
Average turnround time	Hours											42	42

Table A. 8   Selected performance indicators, Port Hedland Port Authority

Financial indicators	Units	1993 – 1994	1994 – 1995	1995 – 1996	1996 – 1997	1997 – 1998	1998 – 1999	1999 – 2000	2000 – 2001	2001 – 2002	2002 – 2003	2003 – 2004	2004 – 2005
Return on assets	%	2.0	2.3	2.9	2.1	1.3	1.2	1.2	−324	6.0	6.6	4.7	4.8
Dividend payout ratio	%	7.7	6.7	18.6	36.4	21.4	55.6	28.6	−0.3	50.0	50.0	96.4	50.7
Current ratio	%	4.6	5.3	4.9	5.7	7.1	6.3	6.9	3.3	378.5	285.7	256.9	100.3
Debt/equity	%	3.3	3.1	2.1	1.9	1.8	0.0	0.0	0.0	0.0	0.0	0.0	0.0
Operating profit (before tax)	$/million	2.6	3.0	4.3	3.4	1.9	2.0	2.2	−137.3	2.6	3.0	2.2	2.4
Port authority costs/unit of cargo	$/Tonne	0.15	0.14	0.14	0.14	0.15	0.14	0.16	0.19	0.14	0.15	0.15	0.16
Port authority revenue/unit of cargo	$/Tonne	0.2	0.19	0.21	0.19	0.18	0.17	0.19	0.19	0.18	0.19	0.19	0.18
Non-financial indicators													
Total cargo throughput	million tonnes	53.3	60.3	63.9	68.3	69.8	67.2	65.4	72.9	72.4	81.4	89.8	108.5
Containerised cargo	'000 TEUs												
Average turnaround time	Hours												
Average total employment					18				17	17	18	18	21

APPENDIX B

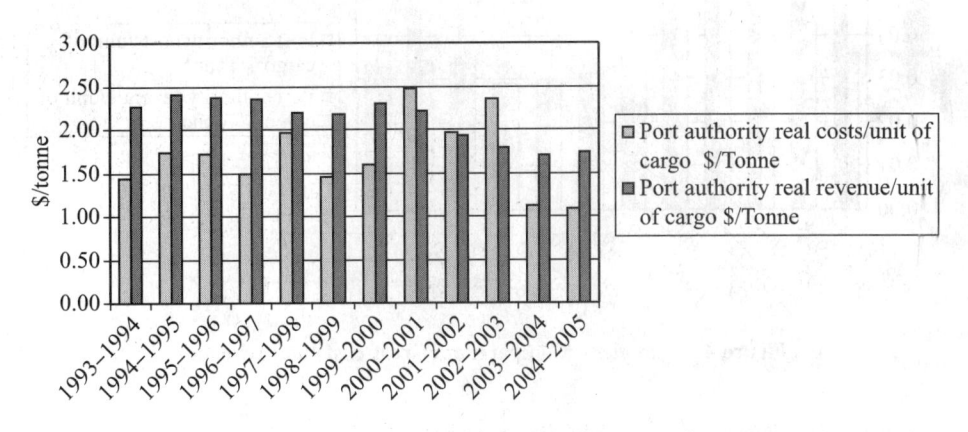

Figure 1　Albany：Comparison of real costs and revenue

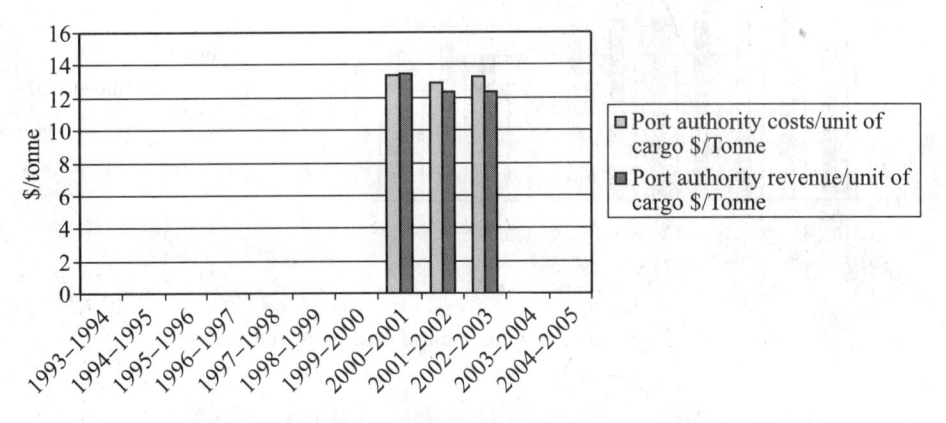

Figure 2　Broome：Comparison of real costs and revenue

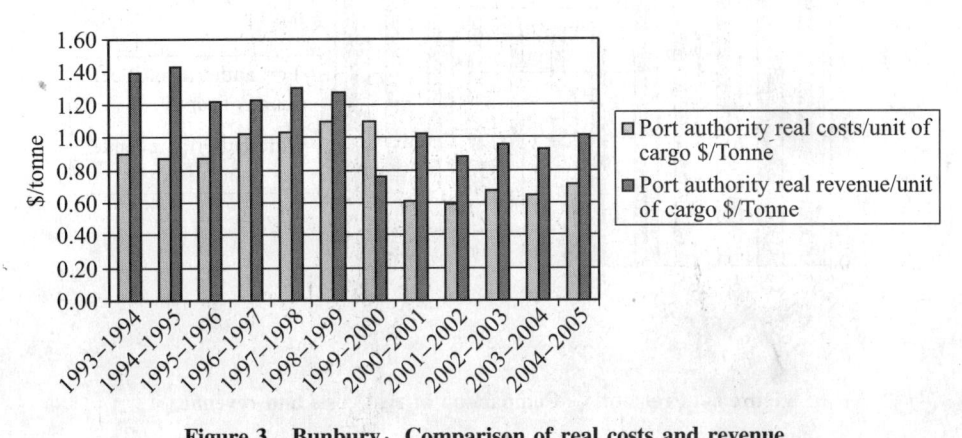

Figure 3　Bunbury：Comparison of real costs and revenue

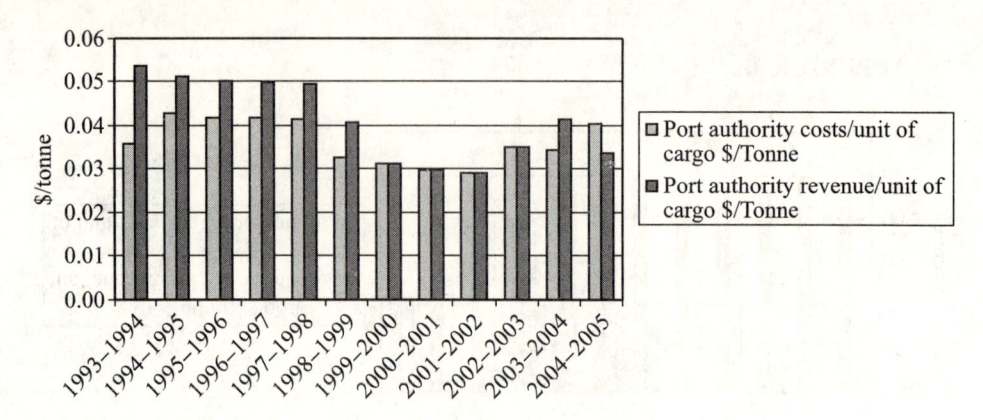

Figure 4　Dampier：Comparison of real costs and revenue

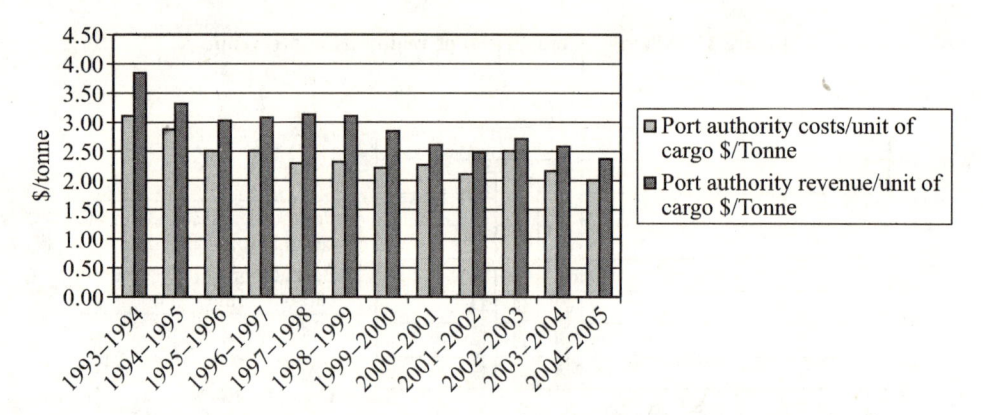

Figure 5　Esperance：Comparison of real costs and revenue

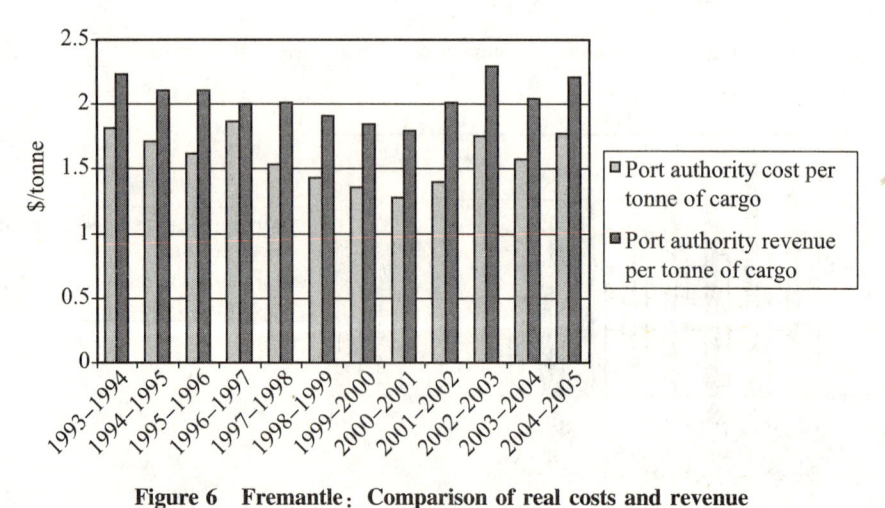

Figure 6　Fremantle：Comparison of real costs and revenue

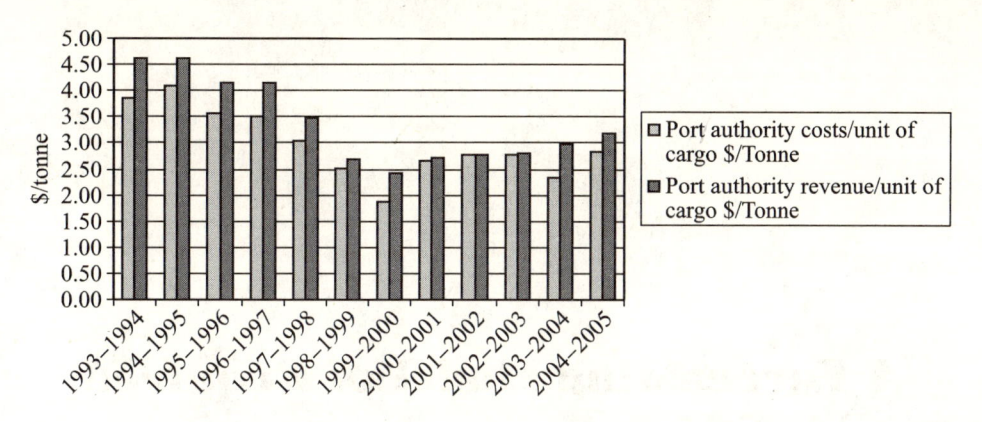

Figure 7　Geraldton: Comparison of real costs and revenue

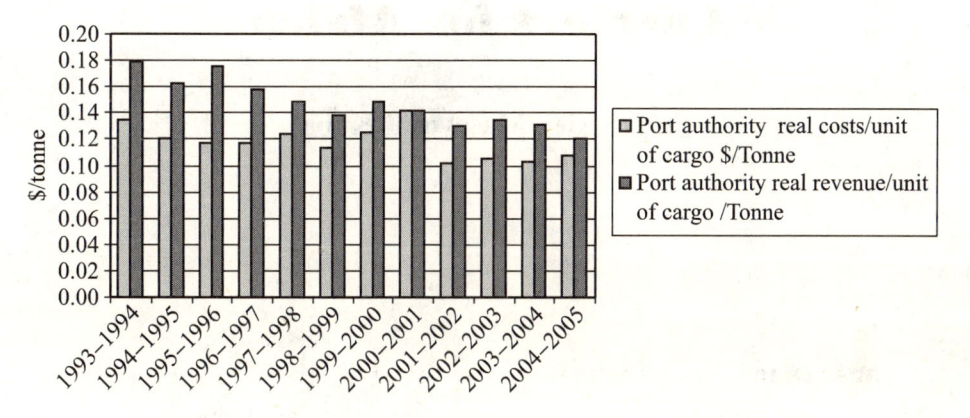

Figure 8　Port Hedland: Comparison of real costs and revenue

A Forecasting Model of Required Number of Wheat Bulk Carriers for Africa

Jingci Xie, Masayoshi Kubo[*]

[**Abstract**] The ocean transportation of grain bulk carriers is promoted by development of ocean economic. With the development of coastal region, the cargo transportation will become more and more important, especially for the resource such as grain, oil and coal. In this study, a model is built to estimate the number of grain bulk carriers needed for wheat based upon analyzing the relationships between Tons and Ton-miles of Africa wheat transportation. We find that the agricultural policies greatly affect the wheat transportation to Africa. Then, using two scenarios, we predict how many ships are necessary for the maritime transportation of wheat from other places to Africa in the future. We believe that this research is extremely useful to maritime transportation of wheat to Africa.

---

[*] Jingci XIE, Division of Maritime Logistics Sciences, Graduate School of Maritime Sciences, Kobe University, Minami 5-1-1, Fukae, Higashi-Nada, Kobe 658-0022. Japan. E-mail: 048d847n@ stu. kobe-u. ac. jp. Tel: 0081-78-431-6339, Fax: 0081-78-431-6364.

Masayoshi Kubo, Professor in Division of Maritime Logistics Sciences, Graduate School of Maritime Sciences, Faculty of Maritime Sciences, Kobe University, Minami 5-1-1, Fukae, Higashi-Nada, Kobe 658-0022. Japan, E-mail: kubomasa@ maritime. kobe-u. ac. jp, Tel: 0081-78-431-6339, Fax: 0081-78-431-6364.

【Key Words】 Maritime transportation, coastal region development, regression model, trade, bulk carriers

1. Introduction

Wheat is very important to people as a food. Especially in Africa, the present agricultural condition is dangerous. Starvation frequently happens in many countries. The cause of starvation is often considered to be the explosive growth of the population in Africa. However, climatic aberration, civil war, the depression of national economies, and the failure of economic policy in the world economy system are also important factors. In other words, the cause is composed of various complex factors, such as natural, social, and political factors. As a result, food shortages are very serious for the African people.

In addition, the distance between producers and consumers becomes longer than before in accordance with the expansion of international food trade. As a result, distribution channels are becoming complicated, and consumers do not know where food comes from and how it is produced. Algela Paxton (Algela 1994) argues about the danger of the long-distance transportation of food.

However, little research has been done on the maritime transportation of wheat in specific regions of the world. Xie and Kubo (Xie and Kubo 2007a) discussed the maritime transportation system of wheat, which includes the maritime tons, ton-miles, and value. In this study, a model is built to estimate the number of grain bulk carriers needed for wheat based upon analyzing the relationships between Tons and Ton-miles of Africa wheat transportation. We find that the agricultural policies greatly affect the wheat transportation to Africa. Then, using two scenarios, we predict how many ships are necessary for the maritime transportation of wheat from other places to Africa in the future. We believe that this research is extremely useful to maritime transportation of wheat to Africa.

2. Wheat Trade Flow of Important African Countries

Now, the international wheat market is an oligopoly, which includes the United

States, Canada, the EU, Australia, and Argentina. These countries hold 85% of the trade share. On the other hand, Table 1 shows the seaborne trade of main African countries. The share of these countries is more than 73% in Africa. Egypt, Algeria, and Morocco are the top three importing countries in Africa. The amount of import of the three countries was 7.76 million tons and the total share of the three countries was 66% in 1980. They became 13.7 million tons and 57% in 2000, respectively. The total share of the three countries occupied about half of African wheat marketing. The other four secondary countries are Nigeria, Tunisia, Sudan, and Ethiopia. The total import of all seven countries was 9.67 million tons, and the total share of the seven countries was 82% in 1980. They became 19.47 million tons and 81% in 2000. The total share of the seven countries is about 80% of the African wheat market.

The trade of wheat is analyzed by using FAO trade data (FAO 2001). Africa imports wheat from Europe, North and Central America, South America, and Oceania. Until now, 49 countries import wheat from 15 other countries, including three North and Central American countries, nine European countries, two South American countries, and one country in Oceania.

Here, we mainly analyze the African trade flow using the panel data of the top three countries. Panel data sets are two-dimensional, whereas time series and cross-sectional data are both one-dimensional. The cross-sectional data are the O-D data of wheat trade. O-D data is Origin and Destination date. Cargo movement is usually represented by origin-destination (OD) table or matrix. By analyzing the O-D date, the trade flow of wheat can be expressed clearly. Here, the O-D data of 1998 of wheat trade are analyzed. Time series data shows the volume of imports from the partner countries for three countries from 1980 to 1998.

In 1998 the O-D data of three countries' wheat trade are shown in Table 2. The trading partners of Egypt are the EU, USA, Australia, etc. The volume of import is almost from the EU, USA, and Australia. Based on the share of trading partners, the United States occupied 63%, and EU and Australia occupied 15% and 20%, respectively. The trading partners of Algeria are the EU, Canada, the United States, etc. Three countries occupied 96% of Algeria's importing market. The trading partners of Morocco are the EU, Canada, Hungary, USA, etc. Based on the share of the

Table 1 Seaborne Trade of Important African Countries

Country	1980 Import	1980 Share	1985 Import	1985 Share	1990 Import	1990 Share	1995 Import	1995 Share	2000 Import	2000 Share	2004 Import	2004 Share
Egypt	4.42	38%	4.52	31%	5.40	37%	5.07	29%	4.90	20%	4.37	18%
Algeria	1.70	14%	3.03	21%	2.61	18%	3.50	20%	5.37	22%	5.03	21%
Morocco	1.65	14%	1.92	13%	1.36	9%	2.55	14%	3.44	14%	2.65	11%
Total thtee	**7.76**	**66%**	**9.48**	**66%**	**9.37**	**65%**	**11.12**	**63%**	**13.70**	**57%**	**12.05**	**50%**
Nigeria	1.10	9%	1.43	10%	0.03	0%	0.61	3%	2.22	9%	2.61	11%
Tunisia	0.65	6%	0.49	3%	0.90	6%	1.65	9%	1.39	6%	1.04	4%
Sudan	0.16	1%	0.58	4%	0.50	3%	0.11	1%	1.00	4%	1.22	5%
Ethiopia	0.00	0%	0.00	0%	0.00	0%	0.51	3%	1.16	5%	0.87	4%
Total seven	**9.67**	**82%**	**11.97**	**83%**	**10.80**	**75%**	**14.00**	**80%**	**19.47**	**81%**	**17.79**	**74%**

Table 2 Trade Flow of Egypt, Algeria, and Morocco in 1998 (Share)

Africa\Important importing country	EU	USA	Australia	Canada	Hungary	Total 4
Egypt	15%	63%	20%	0%	0%	98%
Algeria	43%	13%	0%	40%	0%	96%
Morocco	50%	9%	2%	19%	11%	91%

37

trading partners, the EU occupied 50%, USA occupied 9%, and, Canada, Australia, and Hungary occupied 2%, 11%, and 19%, respectively.

Figure 1 shows the changes in exports to Africa. As shown from (a) to (c) in Figure 1, the volume of wheat import increased in Egypt, Algeria, and Morocco according to the panel date from 1980 to 1998. The EU increased the exports to the three countries from 1989. Canada is also in the expansion of exporting to Algeria and Morocco. The condition of Australia's exporting to Egypt is almost stable. In the case of the USA, the volume of exports to Egypt is increasing. However, after the volume of exports to Algeria and Morocco peaked in the 1980s, the exports declined. Therefore, there is a competitive relationship among the EU, Canada, and the USA. Figure 1 (d) shows changes in import amounts (three countries). As shown in Figure 1 (d), in the case of Egypt, the imported volumes of wheat peaked in 1974, 1981, 1989, and 1996. In the case of Algeria, the imported volumes of wheat peaked in 1975, 1985, 1989, and 2004. Finally, the imported volumes of Morocco peaked in 1975, 1985, 1989, and 2004. It clearly shows the increasing tendency of three countries' import values.

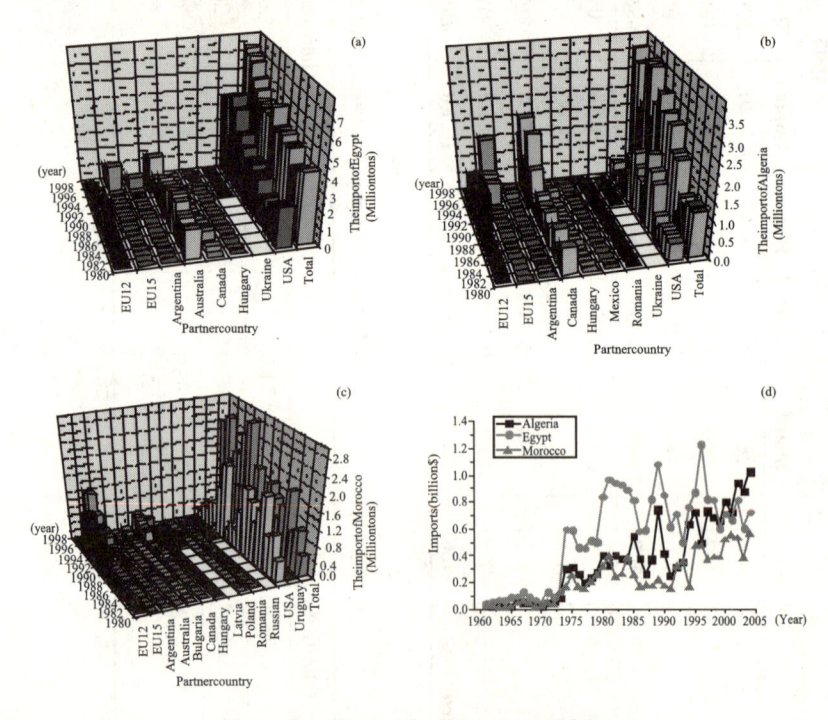

Figure 1　Change in Exports to Africa

3. Forecasting the Required Number of Wheat Bulk Carriers

Forecasting the required number of wheat bulk carriers is difficult at the country level because the changing tendency of the import volume is very sharp. Therefore, in this research, we predict the import volume at the continental level. According to this method, compared to the cases of the country level, the import volume becomes very flat and smooth, so, the function system can be applied, and the accuracy can be raised (Kubo 2004, 2006).

3.1 Examination of Wheat Bulk Carrier Size and Depth of Water

Grain transportation is usually performed by maritime transportation. The means of transport is grain bulk carriers, in which, the Panamax Ship is the dominant type of all ships. The Ship of Panamax type is the largest ship which can pass through the Panama Canal, and its loading capacity is about 70,000 tons or less. It is clear that 10,000 ~ 60,000 DWT bulk carriers are often used. The average depth of each strait on the navigation route is 20 meters or more; therefore, it appears that bulk carriers from 10,000 to 60,000 DWT could navigate the waters. The depth of importing and exporting ports is 10 meters and greater; therefore, 60,000 DWT bulk carriers are suitable for shipping and unloading wheat at ports.

3.2 Forecasting Model for Required Number of Wheat Bulk Carriers

Here, the volume of imports will be the volume of the maritime transportation because the whole wheat will be shipped. In the following analysis, the navigation days of a vessel are considered as 95% of 365. The navigation days, annual voyages, and necessary ships in the export navigation route from each exporter to an African importer could be determined by using the following equations (Kubo, 2005):

$$(1) \quad D_{ij} = \frac{d_{ij}}{v \times 24} \times 2 + D_{ED} + D_{LU}$$

$$(2) \quad N_{ij} = \frac{365 \times 0.95}{D_{ij}}$$

$$(3) \quad TS_{ij} = \frac{W_{ij}}{DWT \times N_{ij}} = \frac{W_{ij}d_{ij} + W_{ij} \times 72v}{DWT \times 4161v} = \frac{W_{ij}d_{ij} + W_{ij}A}{B}$$

$$(4) \quad TS_{\text{North \& Central America} \rightarrow \text{Africa}} = \frac{1}{B}\sum_{i=1}^{3}\left(\sum_{j=1}^{49} W_{ij}d_{ij} + A\sum_{j=1}^{49} W_{ij}\right)$$

$$TS_{\text{Europe} \rightarrow \text{Africa}} = \frac{1}{B}\sum_{i=4}^{12}\left(\sum_{j=1}^{49} W_{ij}d_{ij} + A\sum_{j=1}^{49} W_{ij}\right)$$

$$TS_{\text{South America} \rightarrow \text{Africa}} = \frac{1}{B}\sum_{i=13}^{14}\left(\sum_{j=1}^{49} W_{ij}d_{ij} + A\sum_{j=1}^{49} W_{ij}\right)$$

$$TS_{\text{Oceania} \rightarrow \text{Africa}} = \frac{1}{B}\sum_{i=15}^{15}\left(\sum_{j=1}^{49} W_{ij}d_{ij} + A\sum_{j=1}^{49} W_{ij}\right)$$

$$TS_{\text{Africa}} = TS_{\text{North \& Central America} \rightarrow \text{Africa}} + TS_{\text{Europe} \rightarrow \text{Africa}} + TS_{\text{South America} \rightarrow \text{Africa}} + TS_{\text{Oceania} \rightarrow \text{Africa}}$$

Here, $A = 72v$, $B = DWT \times 4161v$, i: exporter, j: importer. D_{ij} is the number of days necessary for a round trip from i to j; d_{ij} is the distance from i to j (miles); v is the speed of the grain bulk carrier (miles/hour); D_{ED} is the number of required days for entering and departing ports in one round trip (3 days); D_{LU} is the number of days required for loading and unloading in ports in one round trip (3 days); N_{ij} is the number of annual voyages of i and j; TS_{ij} is the number of ships for i and j; W_{ij} is the seaborne trade from i to j (tons); DWT is the dead weight tonnage of a grain bulk carrier. Based on this model, the required number of grain bulk carriers can be predicted based on the maritime tons and maritime ton-miles.

3.3 Import Analysis of Africa

Figure 2 shows the analysis of imports in Africa. As shown in Figure 2, 1986, 1988, and 1995 are very important years because the annual tendency of trade volume changed most in these years according to the regression equations of four areas. In the following part, we will make a detailed analysis.

In the case of 1986, the United States applied the Export Enhancement Program (EEP) as retaliatory measures against the wheat export of the EC using a subsidy. At the same time, Australia and the other 14 so-called fair exporting countries gathered together in Cairns City, Australia. They protested against such unfair trade, especially aimed at the export subsidy because they were suffering from the falling wheat prices.

Hence, these countries were very seriously affected by the export volume after 1986.

In 1988, crops suffered due to the drought in North America. Therefore, wheat output decreased greatly in the North America and Central America regions, and the exports decreased. On the other hand, wheat stock of the EU was increasing and was recorded at 11,300,000t in 1989. Europe got its best chance to enhance the exports to Africa in that time.

Since the 1980's, the government of the EU has always paid a subsidy to protect domestic agriculture. Consequently, the quantity of production increased continuously and became overproduction. In order to solve this problem, the EU carried out the dumping export using the export subsidy. The result is that import demand was slumping in the developing countries, and the wheat price was dropping; moreover, the export competitiveness of traditional exporting countries, such as Canada, Australia, and Argentina, became remarkably weak. In this situation, the world's largest principal exporting nations, United States and the EU, began dumping exports. Because subsidies gradually create a trade barrier, the General Agreement on Tariffs and Trade (GATT) was established. The policies that affected agricultural trade were subjected to systematic multilateral controls for the first time by the Uruguay Round's 1994 Agreement on Agriculture (AoA). The negotiation resulted in an agricultural agreement, and the contracting states were obliged to reduce the subsidy to agricultural producers.

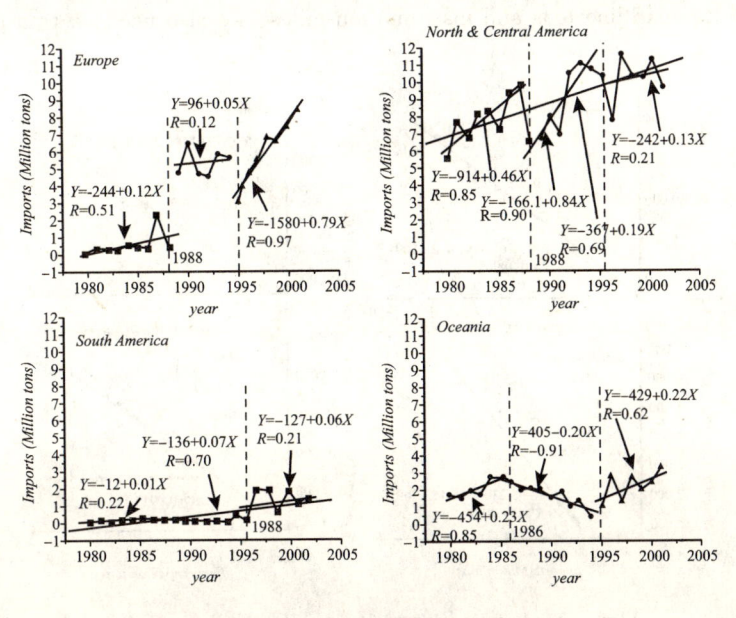

Figure 2　Import Analysis of Africa

From 1995, the GATT became the WTO, and the WTO plays the role of an engine for international trade expansion of wheat. Based on the agricultural agreement of WTO, some policies were put into practice, such as the reduction of farm subsidies, the customs duty, and the amount of import quotas. These policies promoted trade liberalization.

Next, we make a prediction for the imports by a straight line approximation for four areas. It is difficult to make a long-term forecast because the agricultural trade is greatly affected by policies. Consequently, we made two scenarios to predict the wheat import in Africa until 2010. In the first scenario, we predict the imports from North and Central America and South America using the data from 1980 to 2001. In addition, the imports from Europe and Oceania are predicted by using the data since 1995. In the second scenario, we predicted all cases by using the data since 1995.

3.4 Relationships between Tons and Ton-miles

Firstly, we select out some ports as suitable export and import ports, and then we can get the navigation distance among them. Since the movement of wheat can be understood from the OD tables, the maritime transportation ton-miles can be calculated (Xie and Kubo 2007 a, b). Because the forecasting model is based on the prediction of the maritime tons and maritime ton-miles, we also need make a prediction

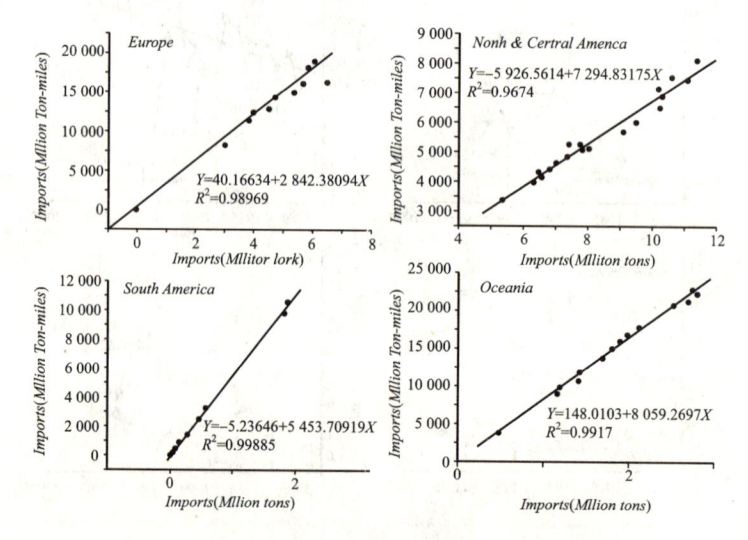

Figure 3   Relationships between Tons and Ton-miles

for the maritime ton-miles. In order to predict the maritime ton-miles, we can predict them from the relation between ton and ton-miles in the Figure 3. As shown in Figure 3, the coefficient of determination is more than 0.96, so strong correlations can be seen. It shows that the partner countries of trade do not change in four areas. It is a perfect condition for sufficient prediction, so the required number of grain bulk carriers can be predicted from the maritime tons and maritime ton-miles.

3.5 Prediction Result

By using 60,000 DWT grain bulk carriers and considering 10 nautical miles/hour, we make a prediction by the above model and get the results as shown in Table 3. In the first scenario, the number of grain bulk carriers will increase to 37, 4, 17, and 22, respectively, from North and Central America, South America, Oceania, and Europe in 2010. Therefore, it is 81 of the total necessary carrier for Africa. In the second scenario, the number of grain bulk carriers will increase to 34, 5, 17, and 22, respectively from North and Central America, South America, Oceania, and Europe in 2010. As a result, it is 78 of the total necessary carrier for Africa.

Table 3　　　　　Predicted Number of Bulk Carriers Needed

year	North & Central America Scenario 1	Scenario 2	South America Scenario 1	Scenario 2	Oceania	Europe	Total Scenario 1	Scenario 2
2002	32	31	3	4	11	13	59	59
2004	33	32	3	4	12	15	65	64
2006	35	33	4	4	14	18	70	68
2008	36	33	4	5	16	20	75	73
2010	37	34	4	5	17	22	81	78

4. Conclusions

We summarize the results as follows:

1) We analyzed the panel data of African trade and found that there is a competitive relationship among the EU, Canada, and the USA.

2) We found a strong correlation between the tons and ton-miles. The information was valuable for predicting the appropriate number of required ships.

3) We built a model to estimate the number of grain bulk carriers needed for wheat.

4) We created two scenarios for predicting the imports in Africa, but could not determine which scenario would actually happen in the future because the agricultural trade is greatly affected by various policies.

References

1. Algela, P., (1994). Food Miles Report: the Dangers of Long Distance Food Transport. The S. A. F. E. Alliance.

2. FAO., (2001). FAO Statistical Databases.

3. Kubo, M, and M. Purevdorj, (2004). "The Future of Rice Production and Consumption." Journal of Food Distribution Research 35 (1): 128-142.

4. M. Purevdorj, and Kubo, M., (2005). "The Future of Rice Production, Consumption and Seaborne Trade: Synthetic Prediction Method." Journal of Food Distribution Research 36 (1): 250-259.

5. Kubo, M, and Xie, Jingci., (2006). "A Continental Method for Estimating the Volumes of Maritime Transportation of Wheat." International Association of Maritime Economists Annual Conference Proceedings. CD-ROM. July.

6. Xie, Jingci, and Kubo, M., (2007a). "A Continental Method for Estimating Wheat Production, Supply, and Maritime Transportation." Journal of Food Distribution Research 38 (1): 182 – 195.

7. Xie, Jingci, and Kubo, M., (2007b). "Research on Wheat Maritime Transportation System in Asia." Journal of Logistics and Shipping Economics, pp. 105 – 114.

Managing Australian Defence Force Activities in Marine Protected Areas: Using Jervis Bay as a Case Study

Lingdi Zhao, Xiaohua Wang, Brian Lees[*]

[**Abstract**] Australian Defence Force has done training activities in marine areas even some marine protected areas for a long time. These activities may cause pollution to the environment and related animals both directly and indirectly. So it is necessary to do some research on the environmental influence of ADF activities and try our best to protect the natural environment. In this essay, we take Jervis Bay Marine Park as a case study to study the methods of environmental management of Australian Defence Force Activities. Through our spot investigation, we found that the ADF has some special power in JBMP and their activities certainly did negative impact on not only the environment but also the surrounding communities. To solve these problems, the common citizens and the authority of ADF must shape a good relationship to reduce misunderstanding and the environmental management in Jervis Bay Marine Park should be increased in the future.

---

[*] Lingdi Zhao, Professorin College of Economics, Ocean Univervitg of China, Qingdao, 266071, China, visiting scholar in The Universitg of Now South Wales at ADFA. E-mail: Lingdizhao512@163.com. Xiaohua Wang, Lectwrer in The Universitg of New South Wales at ADFA. Brian Less, Professor in The Univerity of New South Wases at ADFA.

[Key Words] ADF activities   jervis bay marine park   Environmental Management

JEL Classification: Q25, Q53

1. Background

Jervis Bay Marine Park (35°04′S, 150°44′E) is approximately 180km south of Sydney and 20km southeast of Nowra in the Batemans marine bioregion. It spans over 100km of coastline and adjacent ocean extending from Kinghorn Point in the north to Sussex Inlet in the south and includes most of Jervis Bay. The park covers an area of approximately 21450 hectares. And the NSW Government established the park in 1998.

Jervis Bay was chosen as a case study to encourage discussion about how Defence activities can be successfully conducted with due regard for the values and constraints of marine and terrestrial national parks, and the needs and aspirations of the local community. Jervis Bay is a most complex location in terms of environmental management. Lessons learned there from managing Defence activities in an environmentally sustainable manner will provide important guidance for the sustainable conduct and improvement of ADF training and exercise activity management in other regions.

The citizens living around Jervis Bay Marine Park may complain about the pollution and noise which were brought by the denfence force activities. However, the manager of JBMP said that ADF is very careful about JBMP environment and has looked after the environmental impact through joint work. The researcher from ADF claimed that ADF really worries about the environment of JBMP and every action of ADF before undertaken must apply for ECC (Environment Clearance Certificate) also maybe JBMP will be polluted in 7 years if fishing is still allowed in JBMP. From what has been said by the manager of ADF, we know that 1 kilogram explosive is allowed in JBMP, but must ask for permission in advance. Shipping and diving in JBMP have to apply for ECC and is subject to procedure cards. So according to what he said, under the joint management of ADF and JBMP authority, the environment situation of JBMP will not be polluted in the future and to some extent the activities that the ADF taken in JBMP will do good to the protection of the environment.

Even with rigorous debate and discussion of the above issues, there are still more questions than answers on how Defense activities may impact on the maritime environment in the future and how this can, and should, be dealt with. The issue of protecting the maritime environment is going to become increasingly important for both the RAN and the wider community both nationally and internationally. The insights obtained during this Seminar should therefore be regarded as a starting point for ongoing analysis in the coming years.

From the ACT 1997 Marine parks and the spot investigation of Jervis bay marine park, it is necessary for us to do some research on Australian Defence Force Activities in Jervis bay marine park, and from what we have done, we find that:

1.1 Marine Parks Authority pays more attention to the environment of JBMP

Through the following graph (Figure 1), we can see clearly that the Marine Parks Authority pays more attention to the environment of JBMP and gradually improved their management policies on this marine park.

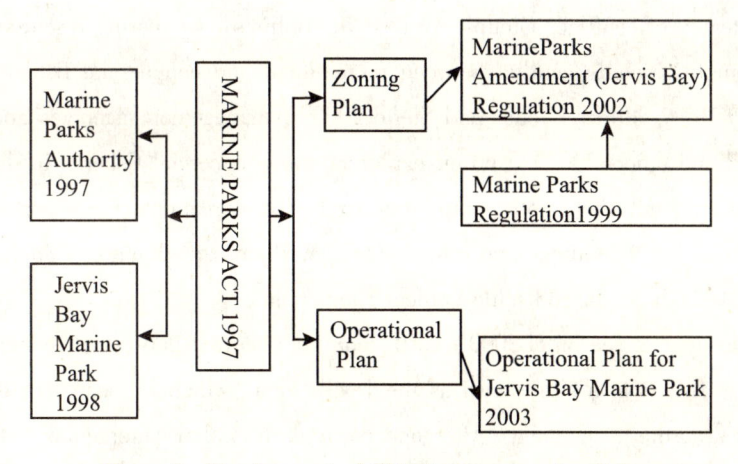

Figure 1　Development of JBMP management policies

Under the guidance of the Marine Parks Act 1997, Marine Parks Authority published Operational Plan for Jervis Bay Marine Park in October 2003. This operational plan outlines the scheme of operations that the Marine Parks Authority intends to undertake or permit in providing for conservation and sustainable use of Jervis Bay Marine Park. One of the most important schemes is the address of specific management areas, which include: habitat and species conservation (Section two); management of activities for ecologically sustainable use, such as fishing, whale watching and scuba diving (Section three); management for Indigenous culture and non-Indigenous heritage (Section four and five); pollution control, management of marine pests, and other management issues (Section six); research, community education, compliance and permits (Sections seven to ten); and the management arrangements with commonwealth government (Section eleven).

The zoning plan which was included in Marine Parks Amendment (Jervis Bay) Regulation 2002 divides the Jervis Bay Marine Park into the various zones and contains special provisions regulating and prohibiting the carrying out of certain activities in those zones. The zoning plan provides for four zones in Jervis Bay marine parks: a sanctuary zone; a habitat protection zone; a general use zone and a special purpose zone, and sets out objects and special provisions applying to those zones.

Before the zoning plan of Jervis Bay Marine Park, there is a conflict in the marine environment between user groups can be categorized into potential, actual, imagined, and philosophical (Orams1999). The response of environmental managers to conflicts, and how best they can be resolved while maximizing ecological benefits, is determined by the category of conflict. Groups involved in conflicts over marine resources are diverse Examples include conflict between recreational and commercial fishers (Ruello and Henry 1977, Murray-Jones and Steffe 2000), management agencies and anglers (Churchill and others 2002), and oil explorers and commercial fishers. In the case of divers and anglers, where both groups seek exclusive recreational use of waters and abundant large fish (Jennings and others 1996, Williams and Polunin 2000), conflict may include both actual and philosophical components.

In the summer season of 2000 – 2001, conflict occurred between divers and shore-based game fishers at a site known as the Docks area, which is located in the lee of Jervis Bay's northern headland. After one particularly violent interaction, where dive boats were attacked with lead sinkers fired by anglers from a high powered slingshot, a dive operator filed a complaint to the police (NSW police report E10957104). A subsequent local newspaper article and editorial gave the angler's viewpoint that the divers had been deliberately scaring the fish away and that some formal delimitation of access rights may be needed (Wright 2001, South Coast Register 2001). Following this incident, the authority identified reduction or elimination of the Docks area conflict as a priority issue.

In addition to the core roles of MPA's protecting biodiversity, habitats, ecosystem processes, and fisheries, conflict resolution may also be attempted (Agardy 2000). However to achieve this, detailed societal data is required (Churchill and others 2002).

To resolve conflict and maximize positive environmental outcomes, a sanctuary zone and no-anchoring zone option in the draft zoning plan was sought to formalize this partition. The human dimension data proved valuable in guiding environmental management in this politically volatile situation.

1.2 ADF has some special power in JBMP

ADF activities in the Jervis Bay region date back to the earliest years of nationhood. The Royal Australian Navy uses Jervis Bay for training, including a small area within the marine park. The Naval College at HMAS Creswell is sited on the southern shore of the bay and inland from HMAS Albatross, the home of naval aviation. The seaward land mass of the bay, Beecroft Peninsula, is one of only three shore bombardment and live areas available to RAN ships in Australia, which is used as well by the other Services.

Offshore in the Tasman Sea, the East Australia Exercise Area (EAXA) is one of the two most important maritime exercise areas in Australia, widely used by the RAAF and the RAN. The value of that training area is accentuated by its proximity to major RAN and RAAF bases in NSW. And it is not just the natural environment in this area that is a challenge for us either. Community and commercial activities in the region are economically and culturally important as well, and do impact upon our activities.

In the operational plan for Jervis Bay Marine Park, 2003, contents 11 emphasized the management arrangements with commonwealth government. From that time on the Marine Parks Authority works in consultation with a number of other Government Departments under a variety of management arrangements, and the details of management actions lie in consulting with Environment Australia and the Department of Defence in accordance with the Management arrangements in place.

1.3 Marine Recreational Activities and Tourism industry are very important for the community of JBMP

Jervis Bay is one of the true gems of the NSW south east region and its value to Australia has been recognized by its National Park status. It's breathtaking beauty, rich cultural and natural significance have long been recognized and appreciated by locals and by domestic and overseas visitors.

This popular marine park caters for a broad range of recreational and commercial activities while conserving marine biodiversity. Recreational activities include fishing, scuba diving, boating, surfing and other beach pursuits including dolphin and whale watching, diving, camping and bushwalking to name just a few. Jervis Bay is reputed to have the clearest waters and whitest sands in not only this country but also in the world.

These beautiful blue waters are home to many dolphins and the opportunity to sight these graceful locals is one of the major features that make this area a magnet for lovers of nature. Whale sighting is also frequent in Jervis Bay and the choices of water sports here are wide. Many people come to dive, fish and explore these famous waters. None go home disappointed.

Also famous at Jervis Bay is the Australian Naval College, HMAS Creswell. This officer training college was established in 1915 and has a rich collection of its history and of model sailing ships.

Towns and villages situated around Jervis Bay include; Culburra Beach, Currarong, Callala Bay and Beach, Myola, Huskisson, Vincentia, Hyams Beach, Green Patch and of course the small township of Jervis Bay itself. There is a large range of accommodation and shopping available amongst these small picturesque townships. Access to the villages on the northern side of Jervis Bay and the southern Shoalhaven Heads is by way of Culburra Beach Rd from the Princes Highway at Nowra.

Culburra Beach, with a population of 3,500 is the regional centre for the coastal villages east of Nowra. Originally designed by Walter Burley Griffin, the architect who designed Canberra, its shops and businesses provide a range of services and places to eat for residents and visitors to the region. The sheltered white sandy beaches of northern Jervis Bay provide safe swimming and excellent fishing. They are especially popular for families with young children or people who prefer still water swimming.

The townships of Callala Bay, Callala Beach and Myola are located on the northern side of beautiful Jervis Bay and are accessed from the Culburra Beach Road. Both towns have a variety of food and other shops. Rental cottages, B&Bs, and caravan parks provide tourist accommodation. So the marine tourism industry plays a very important role in this area, if the number of tourist comes down sharply , the income of these citizens would be decreased obviously. This trend will influence their living standard and make them living a hard life in the future.

2. Impact of ADF activities on environment

Through our investigation, we find that there are two aspects of environmental issues about the impact of ADF activities.

Though the actions that have been taken by ADF seldom cause pollution to the sewage and waste water, the ADF waste such as diesel oil which was embedded in the underground many yeas ago may cause pollution to the groundwater and play a negative effect to the health status of the people who have to contact with or rely on the groundwater surrounding this area during their daily life even a lot of yeas later. Considered the above situation, BNP asked the Defence Force for settlement about this problem, at present ADF is trying their best to deal with the issue and cleaning the polluted groundwater.

The environment issues caused by ADF have certain influence on the main activities that people took before and after JBMP especially on Dolphin watching and whale watching. The details of interaction can be seen from the following table.

Before the establishment of JBMP, in Jervis Bay areas the recreational activities just included diving and angling. And after then, beach-going, kayaking and dolphin and whale watching are the main tour item (Table 1). The most important thing is that on the one hand, diving and angling is declining, and they have no connection with ADF training; on the other hand, dolphin watching and whale watching are increasing, and they have connection with ADF training.

People argued about the influence of ADF training to dolphin and whale watching. The residents of JBMP insist that recent years it is hard for them to see dolphin and whale during their daily life due to the ADF training. And scholars use technical methods to evident what the residents said, they found that although some kinds of sonar beyond of the hearing rang of dolphins and whales, there are still some impacts on them just like the impacts of ultrasonic wave and electromagnetic wave on the human. On the contrary, the research and manager of ADF claims that ADF activities have no impacts on dolphins while some impacts on whales Dolphins were harassed by dolphin watching cruise.

Table 1　　Changing of main activities after the establishment of JBMP

	Main Activities	Connection with ADF training		Reasons
Before JBMP	Diving	No		Boom Industry
	Angling			The third most popular outdoor activity
After JBMP	Beach-going	Uncertain		Uncertain
	Kayaking			
	Dolphin watching and Whale watching	Arguing	Diving declining	Competition from neighboring countries, e.g. Fiji
			Angling declining	(1) Changed structure of angler (2) Zoning
			Residents	No dolphin and whale watch is connected with ADF training
			Scholar	Although some kinds of sonar beyond of the hearing rang of dolphins and whales, there are still some impacts on them just like the impacts of ultrasonic wave and electromagnetic wave on the human

3. Reasons leading to these impacts and necessary policies

The main reasons leading to the above three environmental issues lies in lacking of formal and informal communication between ADF and Community. Nowadays, there are formal and informal communications between the following three sides respectively: residents and ADF; general manager of JBMP and general manager of ADF; senior manager of JBMP and senior manager of ADF, but these communication is obviously insufficient so it should be enhanced in the future through others more useful ways.

In order to solve the above environmental issues, some essential policies must be undertaken:

3.1 It is better to hold more formal stakeholder meetings and organize open day to the common citizens to reduce misunderstanding between different sides and more advice should be given based on the data, experience, and assessment. Through our spot investigation, we have known sometimes there are some misplay during the process of training activities and a serious problem lies in that even the JBMP citizens who lived surrounding the training areas sometimes didn't know the exact timetable and

```
                    ┌─ Residents
                    │    ↕
                    ├─ ADF
              ┌ FORMAL ─ General Manager of JBMP
              │           ↕
              │          General Manager of ADF
              │          Senior Manager of JBMP
              │           ↕
COMMUNICATION─┤          Senior Manager of ADF
              │          Residents
              │           ↕
              │          ADF
              └ INFORMAL─ General Manager of JBMP
                          ↕
                         General Manager of ADF
                         Senior Manager of JBMP
                          ↕
                         Senior Manager of ADF
```

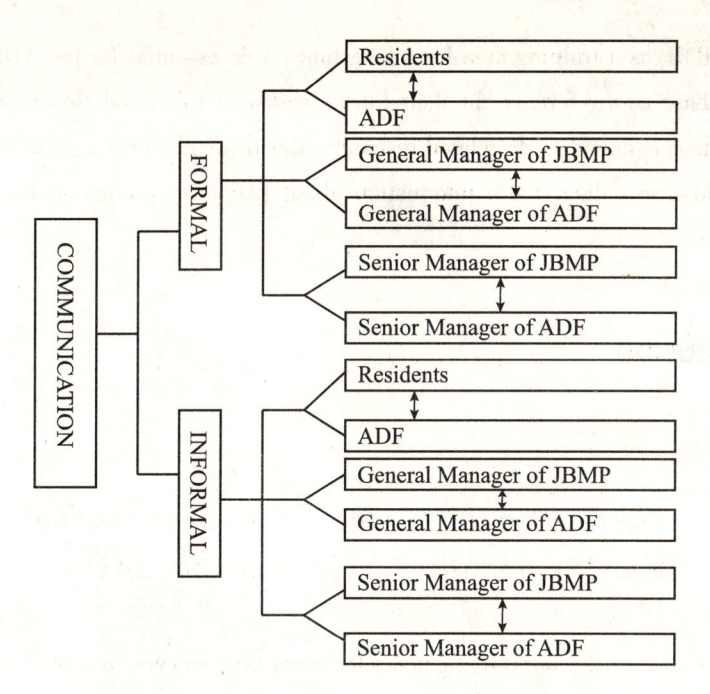

Figure 2 Communication between different groups

content of training activities. So without the effective communication between ADF and common citizens, the citizens may lack of sense of safety, and they may be afraid of being attacked by the misplay of the training and manoeuvre activities. It is really good for the army talking to the community to explain their activity in advance and organize open day to improve public understanding of the Defence Force's efforts. After training, if public still do not feel good or puzzled about some long term impact, the ADF should explain on the newspaper or TV.

3.2 It is necessary to increase environmental management of JBMP. Since now it is unrealistic for the Marine Parks Authority to monitor every single ADF activity and even can not read every application form ADF, the most useful measures to solve this problem according to this policy is increasing environmental managers who have both professional and management knowledge.

3.3 The ADF should set up a database to record the continuous data about the influence that the ADF activities have caused to the environment, especially about the impact on whale and dolphin. And these data must be credible and if the citizens or some related people are suspected about the effectiveness of these data, the ADF could provide evidence to prove the data they used is correct. Since the ADF has used the JBMP as a training area for a long time, it is essential for the ADF to establish a database as a reference for their future work and this database could also provide experts convenient to do related research. Besides collecting a series of data, the ADF should also collect more information about people's opinion on the impact of ADF activities.

4. Conclusion

4.1 There are close relationship among ADF and JBMP, community and JBMP, ADF and community

People all over the world living under the same sky, so every one of us have the responsibility to protect the environment which is now under a serious situation. In order to achieve the same goal, ADF, JBMP, community have already worked together exchanging and discussing the different ideas timely and effectively to avoid misunderstanding. So it is safety for us to conclude that a closer relationship among ADF, JBMP and community will do great help to protect the environment and live harmony in the future.

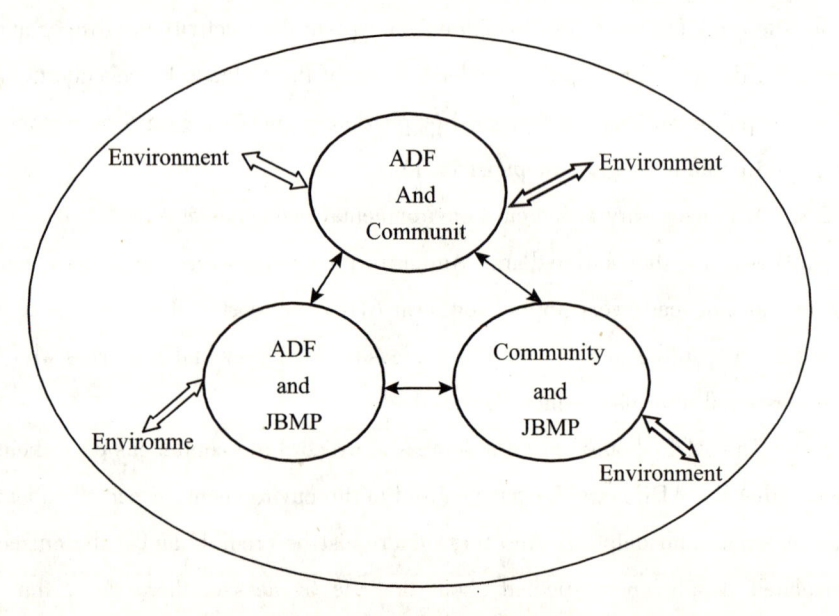

Figure 3   Relationship between diverse groups and environment

4.2 The environment status near the ADF training base has been protected well

During the previous work, through our personally investigation, we found that at the harbor of the ADF training base, there are still many kinds of fish and even the endangered grey nurse shark, so we can believe that now under the joint work of authority of JBMP, ADF and community, the environment of JBMP have been protected pretty good. And some of the citizens also believe that only under further appropriate protection and more attention, the environment of this area could have a better situation. So in the future, it is better for us to pay more attention to enhance on-the-spot investigation and learn more about the different reaction of ADF activities to the short-term and long-term environment.

4.3 The real state value of beachfront will be affected

Now, the activities of ADF has no obvious influence on the real state value of beachfront in Jervis bay, but there is a said that the training area in Sydney will be moved to Jervis bay in the future. Tourist will not be affected by this plan because they will just spend a few days in this area and the activities of ADF will not affect the environment in a short term. However the citizens around Jervis bay complained about this plan, they thought that the ADF activities will have a negative influence not only on the environment but also on the health status of human. So in the future the Jervis bay will not be a suitable place for people to live a comfortable life in their old life times.

According to these consideration, less people would like to choose to buy a house in the area around Jervis bay, so the real state value of beachfront will be decreased and it means that the value of treasure now citizens in areas around Jervis bay will be reduced.

4.4 The issue of dolphin watching and whale watching, and sonar activities may extend and proliferate in the future

(1) From Table 2 and 3, it is reasonable to conclude that compared with the other areas in Australia, JBMP is not a well-known dolphin-watch and whale-watch place.

Table 2 Dolphin-watch and whale-watch place in Australia

	New South Wales	Victoria	South Australia	Western Australia	Queensland	Tasmania	Total
Whale Watching	Cape Byron	Logans Beach	Great Australian Bight Marine Park	Ningaloo Marine Park	Hervey Bay	Freycinet Peninsula	13
	Muttonbird Island Nature Reserve			Pottnest Island			
	Port Stephens		Victor Harbor	Albany	Moreton Island		
	Montague Island Nature Reserve		Point Labatt Conservation Park				
Dolphin Watching	Port Stephens	Melbourne	Sir Joseph Banks Group Conservation Park	Shark Bay	Moreton Island	Freycinet Peninsula	10
			Victor Harbor	Penguin Island		Tenth Island	

Source: Sort out from *Explore Australia*, (new edition) published in Australia in 2003 by Ken.

Table 3 Statistics of dolphin-watch and whale-watch places in Australia

	New South Wales	Victoria	South Australia	Western Australia	Queensland	Tasmania	Total
Whale Watching	4	1	2	3	2	1	13
Dolphin Watching	1	1	3	2	1	2	10

Compare to other dolphin watch place, JB is not well-know. In Explore Australia (new edition, published in 2003), there are 13 Whale Watching places and 9 dolphin watching places, but JBMP not be mentioned. In order to attract more visitors, dolphin must become a trump card. Because whale watch and seal watch are seasonally. Unique to JBMP are the resident pods of Bottlenose dolphins that can be found in the bay all the year round.

(2) USA navy will play an exemplary role

There is a case study about Judge bans Navy from using sonar off Southern California in August 7, 2007. The federal judge (Cooper) in Los Angeles banned the U.S. Navy from using high-powered sonar in nearly a dozen upcoming training exercises off Southern California, ruling Monday that it could "cause irreparable harm to the environment."

Cooper said it was never easy to balance the interests of wildlife with those of national security. But in this case, she said, environmental lawyers have made a persuasive case that the potential harm to whales and other marine life outweighs any harm to the Navy while the court case proceeds. Cooper also ruled against the Navy last year in an earlier case, temporarily blocking the use of active sonar in multinational war games near Hawaii

Over the last decade, scientists have linked mid-frequency active sonar to a number of mass stranding or panicked behavior of whales after naval exercises in the waters off Greece, Hawaii, the Bahamas and elsewhere.

In a well-documented case near the Canary Islands in 2003, an international team of scientists found that at least 10 beaked whales probably succumbed to the bends after bolting to the surface in a panic.

The dead whales washed ashore after the Spanish navy led international military exercises involving warships from the United States and other members of the North Atlantic Treaty Organization. Pathologists found tissue in the whales' internal organs that appeared to have been damaged by compressed inside them.

Ultimately, her decision forced the Navy to negotiate with environmentalists and establish a buffer zone and other precautionary measures before conducting its month long Rim of the Pacific exercises involving 40 surface ships and six submarines from The U.S., Korea, Japan and Australia.

The judge also took issue with an array of measures to protect whales that the Navy has already put in place, including rules that prohibit using the sonar within

1,000 yards of marine mammals. Sound waves may not dissipate levels for more than 5,000 yards, she noted.

So in order to manage Australian Defence Force Activities in Jervis Bay Marine Park, we can take the USA navy as an example. It is unrealistic to ban all the ADF activities in JBMP, but through scientific research we could analysis the impact of these activities and restrict some actions which would bring serious influence on the environment or endangered animals.

References

1. Timp Lynch, Elizabeth Wilkinson, Louise Melling, Rebecca Hamilton, Anne Macready, Sue Feary. Conflict and Impacts of Divers and Anglers in a Marine Park Environmental Management Vol. 33, No. 2, pp. 196–211.

2. Marine Parks ACT 1997.

3. Marine Parks Regulation 1999.

4. Marine Parks Amendment (Jervis Bay) Regulation 2002.

5. Operational Plan for Jervis Bay Marine Park 2003.

6. http://www.jervisbaytourism.com.au.

7. http://www.mpa.nsw.gov.au/jbmp.html.

论海洋发展的基础理论研究

杨国桢　王鹏举[*]

【摘要】 海洋发展基础理论，是指论证人类开发、利用海洋活动的价值、发展道路、社会效益，及其在人类文明中的地位、作用和指导海洋实践的理论。它不是发展理论在海洋区域的直接运用，而是需要以海洋为本位，进行调试和重新设计。这个领域的研究不仅能指导海洋实践，还可以补充和修正现有发展理论甚至人文社会学科的理论缺失，通过理论进展改变人们的观念，为海洋强国作出贡献。

【关键词】 海洋发展　宏观基础理论　应用基础理论

海洋发展是 21 世纪一个具有全局性、战略性和前瞻性的重大理论和实践问题。研究的目标是解决人类围绕海洋进行社会活动过程中所出现的各种矛盾和冲突，以实现海洋开发、利用、管理和保护事业的可持续发展。海洋发展研究的基本属性是人文社会科学，又与海洋科学、海洋技术发展研究互相渗透、交叉，是综合性、边缘性的新兴学科领域。

世界海洋开发与海洋事务的扩展，改变着海洋文明的形态和前进方式。对此进行适时的总结，需要理论的指导和规范。无可讳言，以往海洋发展研究的理论，基本上还是陆地发展理论的延伸，已经难以适应现代海洋开发与海洋事务扩展的要求。故此本文围绕海洋发展基础理论研究的问题，做尝试性的探讨。

---

[*] 杨国桢，中国海洋大学海洋发展研究院学术委员会主任、厦门大学人文学院教授、博士生导师；王鹏举，厦门大学人文学院历史系博士生。

一、构建海洋发展理论的理由

海洋发展指人类通过直接或间接地开发、利用海洋实践活动,影响社会变迁的行为。[1]这个概念是中性的,而海洋发展事实呈现的结果则可能是正面的,也可能是负面的。对海洋发展事实的认识如果仅仅停留在感性层面,海洋发展自然而然也无法成为理论问题进入人们的研究视野,因此,必须通过理论构建将之提升到理性认识的高度。

步入21世纪之后,"海洋的本质和复杂性已使海洋科学发展成一种系统科学,包括了所有的自然科学。最近,社会科学也被包含在内。"[2]错综复杂的海洋问题不仅向海洋科学界,同时也向人文社会学科提出了更高的要求。因此,建设人文社会学科的海洋基础理论无疑具有极大的理论价值和现实意义。

海洋发展基础理论,指论证人类开发利用海洋活动的价值、发展道路、社会效益,及其在人类文明中的地位和作用,指导海洋实践的理论。

构建海洋发展基础理论的客观依据,是自然海洋和人文海洋区分的现实存在。自然海洋指海洋水体(含岛礁、底土、周边海岸带)及其上空的自然组合,是人类海洋发展的环境条件。自然海洋内部存在密切的相互作用,局部扰动可能诱发区域的整体变化,并深刻影响大气的组成和全球气候的走向。人文海洋指人类以自然海洋为基点开展活动的行为模式、生活方式及交往方式,即以"海洋作为其历史生存的空间,并尝试从海洋而非陆地的视角来安排这个世界"[3]的实践产物。海洋发展的地理空间分布受到自然条件和人文条件的共同约束,《联合国海洋法公约》对海洋区域的划分,便与自然状态不完全吻合,包含了沿海国家、岛国与内陆国家之间的利益分配因素,可视为自然海洋与人文海洋两个分类向度的混合产物。

自然海洋约束人文海洋,人文海洋影响自然海洋。海洋资源特别是海底油气、矿产的开发,取决于海洋科学和海洋技术的进展,又深受国内外海洋政治、经济、社会、文化环境和形势的制约。如何维护我国的海洋权益,营建和平发展之海,有赖于综合国力的提高,海洋政治、外交的突破,海洋发展理论和海洋人文精神的支撑,这就提出了综合、宏观研究海洋发展的理论要求。

海洋发展基础理论在学理上是发展理论的分支,但实际情况又远非如此。发展理论源于西方。18世纪中叶,发展(英语development,法语developpent,德语entwicklung)一词被用于描述社会变化的过程。伴随工业革命和进化论兴起,发展的含义主要转向生产和物质生活的进步。第二次世界大战以后,"旨在解释错综复杂的发展过程中的运行规律"[4]的发展研究（development studies）开始兴起。它最初集中于研究发展中国家如何摆脱不发达状态的问题，后来扩大到发达国家的发展史和人类社会整体性的可持续发展问题。

发展研究首先在经济学领域中兴起，形成强调经济增长的发展经济学。它着重"研究经济落后国家或农业国家实现工业化、现代化，实现经济起飞和经济发展"，[5]注重考察"经济所能达到的平均水平是如何影响发展问题的"和"在国家或地区内的居民之间或国家之间的经济分配问题是如何影响发展的"。[6]

其后社会学、政治学等学科对发展问题的研究相继跟进，并形成各自的研究范畴和研究模式：发展社会学是"发展学和社会学交叉形成的一门边缘学科，是从社会学角度对发展的探讨"，[7]主要从社会的层面探讨"现代化的基本特征"、"发展与现代化的模式"和"现代化的进程"，[8]强调通过"社会指标法、历史比较法、政策法等"[8]方法展开研究；发展政治学"以政治发展问题为研究主题，围绕发展中国家如何实现民主转型、如何实现社会稳定、如何克服政治腐败、如何化解政治危机、怎样完成政治文化改造等核心问题而展开学术讨论"；[9]发展人类学由"不同民族集团在经济发展中的地位、作用和收益"的差异出发，"从民族文化的全方位角度进行发展的研究"；[10]"发展哲学是理论化、系统化的发展观，……是经济、政治、文化、生态及人的发展知识的概括总结"，[11]提倡以人为本的新发展观，致力于"为现实的社会发展实践提供价值目标和意义支撑"。[11]

20世纪80年代开始，发展研究理论取得进一步突破，超越单纯的增长，确立了可持续发展（sustainable development）的概念。至今，可持续发展研究蔚成当代发展研究的主流。

发展研究的展开，经历了从边缘（发展中国家）到核心（发达国家），从经济到政治、社会、文化，从理论探索到实践应用的逐渐深化过程。然而，考察各种类型的发展研究论著，可以发现，虽然有个别学者的著作涉及海洋发展在近代经济发展的作用问题，如：奥康内尔指出，"探险家扬帆出海去寻找新大陆"的活动作为西方现代性形成过程中的一部分，"融入了西方的现代化中"；[12]Acemoglu等认为欧洲兴起和大西洋贸易有关，传统现代化理论强调的欧洲特性如新教伦理、国家制度等，不足以充分解释1500~1850年间西北欧地区的兴起，必须综合考虑外生性的濒大西洋自然禀赋条件和内生性的非绝对主义制度。[13]但从总体上来说，海洋发展因素在现有发展理论中仍旧处于边缘地位。其直接后果就是发展研究的构架基本停留在陆地发展上，单纯强调农业社会到工业社会的转型，没有为新的海洋时代提供理论准备。

我国发展研究系从西方引进，基础理论研究薄弱，长期以译介、吸收、综述为主，并多强调应用于中国实践问题的研究，同样很少考虑或不考虑海洋的因素。比如，我国可持续发展研究普遍以大陆为系统，以致"世界地缘环境分布图"[14]也把海洋处理为一片空白，而"我国可持续发展能力资产负债表"[15]也未把海洋统计在内。

由此可见，现有的发展研究理论不能在海洋发展领域直接运用，需要以海洋为本位，进行调试和重新设计。

二、海洋发展的宏观基础理论

海洋发展基础理论不是发展理论在海洋区域的直接运用，应是研究海洋文明自身发展和海洋视角下人类社会变迁的综合理论。鉴于研究对象的复杂性和和变化性，以及海洋人文社会学科尚无范式的状况，海洋发展基础理论需要综合不同学科的视角，选择运用具有针对性的方法和工具，注重各种理念和思想的整合，兼顾不同群体的立场，搭建连接彼此的纽带和桥梁，寻求关于概念内涵的共识，为海洋政策的制定和海洋开发、利用、管理、保护的实践工作提供规范性的指导。

海洋发展基础理论可分为宏观基础理论和应用基础理论两部分。海洋发展的宏观基础理论，旨在寻求良性海洋发展的坚实基础，从抽象、理性的高度总结海洋文明的发展经验，重新审视人们的海洋发展观，提出有效指导海洋实践的根本概念。

从研究内容来说，海洋发展宏观基础理论涉及传统海洋文明和现代海洋文明积累起来的一切海洋知识。在科学技术日新月异、信息大爆炸的今天，知识的进展和观念的变迁日渐融合，海洋知识和海洋观念也基本合二为一。进一步看，海洋知识又存在专业海洋知识和一般海洋认识的区别：前者包括海洋科学技术及人文社会学科中有关海洋的专门知识；后者主要指人们对海洋生产、生活方式及其意义的认知。它们是"社会成员平均具有的信仰和感情的综合，构成他们那自身明确的生活体系"，[16]"形成和完善的过程非常缓慢"[16]的集体意识的一部分。这些知识异常庞杂，甚至常常是相互矛盾的，宏观基础理论研究的要点就在于从中找出连接各种海洋知识的契合点，进行概念内涵的整合。

从方法论而言，海洋发展宏观基础理论要把海洋文明的经验提升为抽象理论，必须考察发展理论和海洋人文社会学科相关理论的学术源流，从中寻找出该理论的海洋经验渊源，以便吸收和再创造；或者采取知识学的路径，对相关理论本身展开话语或谱系的解析，进行语境分析，辨别出体现海洋因素的部分，并在应用于海洋领域前预先调适。

从表面上看，这种理论研究与海洋科技研究、海洋事务没有多大关系，似乎可有可无，实则不然。在实践中，往往可能因为一个基本概念内涵的重新界定，通过理论的进展促使人们趋近问题的本质，改变思维方式和行动方式，进而取得新的突破或解决长期存在的问题。先以海洋渔业（含近岸、近海水产养殖业）的概念为例：我国将其置于大农业的概念之内，称之为"海洋农业"、"海水种养业"、"海上田园"，归属于农业部门管理，在学理上并不科学；如果重加理论建构，就能引起陆地思维方式向海洋思维方式的转换，从而推进管理方式和制度的改革。另外，在涉海活动中，我国存在过于看重眼前经济利益，迷信工程技术至上论信条的问题，常常不能因地制宜地考虑海洋区域的自然和人文特性，进而形成严重的环境、社会负面效应；若能重新改进概念和理论，就能促进人们转变观念，重新认识海洋，纠正偏差。

在自然海洋与人文海洋之间，基本概念需要反思的还有不少，但由于各种海洋专业知识已经学科化，加之研究者本身学术背景的多样性，组织这种对话、沟通将是个漫长的过程。首先，需要海洋科学和海洋人文学科通力合作，摒弃"两种文化"分裂造成的断裂状态，共同建设大海洋学科体系；其次，随着人类开发、利用海洋步伐的加大和科学技术的发展，人文海洋大大拓展，人类对自然海洋的影响日益明显，自然海洋对人类的反作用也越来越强烈，因而亟须海洋可持续发展理论推陈出新；再者，在"大科学"时代的今天，海洋发展与海洋技术条件已密不可分，其他领域的专门知识及源自陆地的各种技术势必大量扩散到海洋领域，这就要求技术转移和发展的过程必须关注自然海洋的特性，高度重视人文海洋特殊的社会条件。

此外，良性的海洋发展还需要政府、学术界、产业界和公众四方的协同努力。一方面，设计出政府、学术界和产业界三方能够共赢的制度安排，实现多层次的海洋可持续发展；另一方面，提高各界的海洋知识水平和海洋事务意识，尤其需要改进公共教育并加强舆论关注，透过常规教育和公共媒体提高国民的海洋意识，改变人们海洋认识水平往往滞后于客观形势要求的状况。

对于我国而言，进行海洋发展基础理论概念磨合的工作尚未全面启动。笔者尝试对海洋发展涉及的几个最基本的概念，如"海洋区域"、"海洋经济"、"海洋社会"、"海洋文化"，作了多学科概念磨合的诠释。[17]今后还要由此进一步扩大，进行全面的清理和讨论，这是很有意义的。

这里必须指出的是：我国目前亟须处理"海洋国家"的概念磨合及本土适用问题，即中国是海洋国家的理论研究。"海洋国家"原是以西方的"海权"理论分析近 500 年来世界性大国崛起现象的一个概念:"作为海洋国家,其牢固的基础是建立在海上贸易之上。"[18] 它诞生于英国维多利亚的自由贸易时代,也是英国工业和经济的优势自由拓展的时代,"自由的海洋与自由的国际市场在自由这个概念中汇合"。[3] 这是对向海外掠夺扩张型海洋发展道路的总结,"海洋国家"因此等同于近代兴起的西方资本主义国家。随着时代的变迁,这种话语受到发展中海洋国家越来越多的质疑和批评。随着海洋资源重要性的日益凸显和《联合国海洋法公约》的落实,"海权"与"海洋国家"的概念内涵不断扩展和外延:"海权"扩大为国家在海洋中的各种实力;"海洋国家"扩大为从传统到当代、以海洋发展为国策、向外用力的沿海国家,包括陆海兼备型国家。在新语境下,"海洋国家"的崛起可以有不同的发展路径。这一话语权的争夺,至今仍很激烈,关系到国家海洋利益的角力。

传统海洋时代,宋、元两朝的中国是大陆国家,又是海洋国家。现代海洋时代,改革开放后的中国,推行海洋强国战略,恢复了海洋国家的地位。我国走的是与西方不同的发展路径,但我们缺少对中国是海洋国家的理论研究,与海洋发达国家,尤其是海洋邻国日本,形成巨大的反差。这种缺失,导致在理论和学术层面上,不能有力地反驳国际上所谓"中国是大陆国家,不是海洋国家"之类否认中国海洋发展合法性的谬论;也不能有力地反驳台湾当局所谓"中国是大陆国家,台湾是海洋国家"的否认中国统一的台独言论;更不能消除国内理论界移植西方理论形成的固定思维方式,特别是由此产生的对中国海洋发展的种种偏见。因此,这是关系中国海洋发展利益的重大理论课题。我们必须积极应对,以创新理论取得话语权,端正世人视听,增进中国海洋发展的软实力。

三、海洋发展的应用基础理论

海洋发展应用基础理论是海洋实践活动的理论反映,强调发现和分析海洋问题,并非既存理论在海洋领域的全盘套用,还要求能够依照海洋自身特性调整和修正有关理论。它包括海洋经济学、海洋管理学、海洋法学、海洋政治学、海洋社会学、海洋军事学、海洋史学、海洋文化学等学科。这些不同的学科拥有相对独立的理论范式,使用不同的模型、方法,从不同的角度或侧面应用于具体的实例,来分析海洋发展具体领域的规律、模式、方法和实现机制等,并为海洋实践活动提供可操作性的理论指导。由于这些学科尚未定型,应用基础理论的研究还有很大的发展空间。

（一）海洋经济学

现有海洋经济学一般被归入应用经济学，或置于区域经济学、产业经济学之下，往往强调将经济学基本原理应用于海洋区域。采用这种强烈且简单的还原论立场，海洋经济势必成为陆地经济的派生。如此一来，海洋自身的特点不能被有效揭示，经济学理论也无法真正步入海洋。

目前中国对海洋经济学的讨论，重点接近于产业经济学，关注"开发利用海洋的各类产业及其经济活动"，以及相关"接口产业及陆海通用设备的制造业"等。[19]在这种情况下，海洋在经济史上的意义被忽略，曾经是近代海洋经济主体并在现代化中发挥作用的海洋商业退出海洋经济学领域，暴露出海洋经济学结构的不合理。

退一步说，海洋经济学只是指当代海洋经济学，也不仅仅是应用经济学、区域经济学、产业经济学理论的应用，还应该探讨知识经济学、空间经济学、新制度经济学、技术经济学等新兴经济学理论在海洋经济领域的理论应用和创新问题。比如，知识经济学"运用经济学方法研究知识价值的产生、转化和应用过程，探索知识的经济属性和商品化途径，确定知识经济价值的标准和方法。"[20]"现代海洋经济是一种知识经济。海洋知识经济就是以知识增长和高技术发展为基础的海洋经济。"[21]它无疑是海洋经济的主导和关键部分。在对海洋知识经济的研究之中，除去利用知识经济学主导模式之外，还应当探讨海洋产业中的知识服务和行业生产统合的应用理论。

经济地理学（空间经济学）的主要研究对象是"经济活动的区位选择及其原因"。[22]经济地理学经典区位模型——杜能模型，以及在其基础上经修正、再创造而产生的城市体系和产业体系理论模型，都是为陆地区域设计的。在研究海洋经济地理特性、探讨海洋产业布局等问题时，仅有陆地特性的区位模型显然不足，无疑需要构建出海洋区域条件下的经济地理区位模型。这方面的应用理论研究成果，对海洋区域规划的科学性也具有指导意义。

目前，海洋在我国是众多污染的最终承受者，由此引起的海洋损失又不反映在绿色 GDP 等统计数据之内，而海洋生态系统的整体性和迁移性，以及区域归属和产权的不确定性，又加大了计量难度。尽快完善统计口径，制定出海洋可持续发展指标体系，应是应用理论研究的课题。

（二）海洋管理学

现代中国的海洋管理学，是从海洋发达国家（主要是美国）引进的。"在海洋事业（含开发、利用、保护、权益、研究等）活动中发生的指挥、协调、控制和执行实施总体过程中所发生的行政与非行政的一般职能，即是海洋管理。"[23] 这要求国家通过政策、法律、经济、行政等手段实现对权益所及的海洋区域内资源开发和人类海上活动的控制，[24] 即以国家作为海洋管理的主体，进行海洋综合管理和海洋区域管理，将海洋管理内容划分为海洋权益管理和海洋经济管理。以这种理论进行海洋管理体制的建设，带有鲜明的时代特色。

各国的海洋管理体制都与其国情和历史传统相关。当代海洋管理理论移植到中国后，产生了与海洋管理现状相矛盾的问题。所谓"群龙闹海"，就是以行业管理为主导。在这种状况下，即使在个别地方政府主导的海洋综合管理改革试点取得成功，也难于全面铺开。过分强调国家行政主体性，无疑是没有看到中国历史上的国家主权有一个从模糊到清晰的过程，忽略了地方、民间层面先于国家层面的海洋开发、利用、管理手段和行为，形成传统并发挥作用。尊重"群龙闹海"的现实存在的合理性，寻求从民间管理、行业管理向综合管理转型的可行路径，是中国海洋管理学应用基础理论研究不能回避的课题。

（三）海洋社会学

海洋社会学的研究刚刚起步，尚未形成理论范式。开展相关的应用理论研究，可以借助发展社会学、环境社会学和海洋经济、海洋管理、海洋文化等学科的理论工具。

我国现在并存着以渔民、渔村、渔业为代表的传统海洋社会，以及以新兴高科技海洋企事业为代表的现代海洋社会。就海洋社会内部而言，传统海洋社会现代化程度低，面临边缘化的处境，采取弃海登陆改行的做法更加剧了这种危机。从理论上探索可操作的参与式发展模式，增进传统海洋社会群体和个人自身发展能力，促进传统海洋社会良性转型，是破解"三渔"难题的需要。现代海洋社会虽处于强势，但多源于陆地，海洋文化积淀不足，在实践中容易沿袭陆地思维，忽略海洋区域的特殊性，采取唯陆地性质的发展策略，例如，在临海工业区填海造地，从长远看将影响海岸带地区和近海的生态环境。如何根据海洋区域的特性，建立现代海洋文化和以海为本的运行机制，也需要应用基础理论的支持。同时，在海洋社会外部，还面临沿海城市加速扩张、沿海人口和社会问题同步递增等问题，使如何协调陆海、共建和谐社会的要求变得非常迫切，因此对应用基础理论研究就显得十分必要。

综上所述，海洋发展的基础理论研究虽然不能立竿见影，取得直接的经济效益，但它是贯彻科学发展观的需要，是海洋实践发展的需要，是建设大海洋学科体系的需要，在海洋发展研究中理应占有一席地位。相信这个领域的研究不仅能指导海洋实践，还可以补充和修正现有发展理论甚至人文社会学科的理论缺失，通过理论进展改变人们的观念，为海洋强国作出贡献。

参考文献

[1] 杨国桢. 人海和谐：新海洋观与21世纪的社会发展 [J]. 厦门大学学报（哲学社会科学版），2005（3）：36.

[2] John G. Field, Gotthilf Hempel, Colin P. Summerhayes. 2020年的海洋科学、发展趋势和可持续发展面临的挑战 [M]. 吴克勤，林宝法，祁冬梅，译. 北京：海洋出版社，2004：153.

[3] 施密特（Carl Schmitt）. 陆地与海洋——古今之法变 [M]. 林国基，周敏，译. 上海：华东师范大学出版社，2006：93，57－58.

[4] 李小云，齐顾波，徐秀丽. 普通发展学 [M]. 北京：社会科学文献出版社，2005：18.

[5] 张培刚. 发展经济学教程 [M]. 北京：经济科学出版社，2001：2.

[6] 德布拉吉·瑞（Debraj Ray）. 发展经济学 [M]. 陶然，等译. 北京：北京大学出版社，2002：7－8.

[7] 张琢，马福云. 发展社会学 [M]. 北京：中国社会科学出版社，2001：21.

[8] 吴忠民，刘祖云. 发展社会学 [M]. 北京：高等教育出版社，2002：16.

[9] 燕继荣. 发展政治学：政治发展研究的概念与理论 [M]. 北京：北京大学出版社，2006：1.

[10] 陈庆德，等. 发展人类学引论 [M]. 昆明：云南大学出版社，2001：47.

[11] 邱耕田. 发展哲学导论 [M]. 北京：中国社会科学出版社，2001：3，29.

[12] 詹姆斯·奥康内尔. 现代化的概念 [A]. 见西里尔·E. 布莱克（Dyril E. Black）. 比较现代化 [C]. 杨豫，陈祖洲，译. 上海：上海译文出版社，1996：24.

[13] Daron Acemoglu, Simon Johnson, James Robinson. The Rise of Europe：Atlantic Trade, Institutional Change, and Economic Growth [J]. The American Economics Review, 2005, 95 (3)：547－579.

[14] 牛文元. 持续发展导论 [M]. 北京：科学出版社, 1994: 21.
[15] 中国科学院可持续发展战略研究组. 2006 中国可持续发展战略报告——建设资源节约型和环境友好型社会 [M]. 北京：科学出版社, 2006: 211.
[16] 埃米尔·涂尔干（Emile Durkheim）. 社会分工论 [M]. 渠东, 译. 北京：三联书店, 2000: 42, 248.
[17] 杨国桢. 论海洋人文社会科学的概念磨合 [J]. 厦门大学学报（哲学社会科学版）, 2000 (1): 95 – 100.
[18] 马汉（Alfred T. Mahan）. 海权对历史的影响 [M]. 安常容, 成忠勤, 译. 北京：解放军出版社, 2006: 68.
[19] 叶向东. 现代海洋经济理论 [M]. 北京：冶金工业出版社, 2006: 8 – 9.
[20] 高洪深. 知识经济学教程 [M]. 北京：中国人民大学出版社, 2006: 55.
[21] 管华诗. 海洋知识经济 [M]. 青岛：青岛海洋大学出版社, 1999: 13.
[22] 藤田昌久, 保罗·克鲁格曼（Paul Krugman）, 安东尼·P. 维纳布尔斯（Anthony J. Venables）. 空间经济学——城市、区域与国际贸易 [M]. 梁琦, 主译. 北京：中国人民大学出版社, 2005: 152 – 153.
[23] 鹿守本. 海洋管理通论 [M]. 北京：海洋出版社, 1997: 49.
[24] J. M. 阿姆斯特朗（Armstrong, J. M.）, P. C. 赖纳（Ryner, P. C.）. 美国海洋管理 [M]. 林宝法, 郭国梁, 吴润华, 译. 北京：海洋出版社, 1986.

Research of the Basic Theories of Ocean Development

Yang Guozhen, Wang Pengju

【Abstract】 The basic theory of ocean development is to instruct the practice of the ocean, which refers to argue the value, paths and social efficiencies of human activity in ocean area, the position and effect of the ocean in human's civilization as well. This theory doesn't mean the direct application of development theories in ocean area, there demands taking the ocean itself as basis, rectifying and redesigning of the theory. Moreover, the research could not only instruct the practice of the ocean, but also could make up and revise the existing development theories, even the theoretical imperfection of humanities and social studies. With the progress of this theory, it can transform the ideas of the people, and then make contribution for an ocean power.

【Key Words】 Ocean development   Macroscopic basic theory   Applied basic theory

JEL Classification: A14

和谐社会建设指向下的政府海洋管理转型

徐质斌[*]

【摘要】 运用和谐社会建设理论观察中国海洋事业,可以多维度地观察到现实中的失谐现象,其根源来自海洋复杂系统自身的脆弱性,更来自社会转型中管理滞后造成的摩擦。新时期海洋工作的基本目标,是追求以社会心理和谐为核心、经济关系和谐为基础、社会群体和谐以及人与自然和谐为突出表现的规范状态。和谐绩效是管理类型的函数。政府海洋管理必须应答和谐社会建设的诉求,实现行政型→治理型、管制型→服务型、权力型→责任型、人治型→法治型、全能型→有限型、经验型→学习型的转变,并建立利益激励、制度约束和公众监督机制予以推进。

【关键词】 和谐社会 海洋管理 转型 机制

中共十七大报告强调"构建社会主义和谐社会是贯穿中国特色社会主义事业全过程的长期历史任务"。从海洋经济领域的实际出发,探讨如何深入落实《中共中央关于构建社会主义和谐社会若干重大问题的决定》,对于海洋事业又好又快地发展,具有重要理论意义和现实意义。本文寻求理论与实践的结合、价值与工具的匹配、诉求与响应的对接,在梳理以往和谐社会研究成果基

---

[*] 徐质斌,广东海洋大学(湛江)海洋经济研究所所长,教授,中国海洋大学海洋发展研究院研究员;邮政编码:524025,Email: qdxzb@sina.com。

础上，根据笔者多年对国家海洋事业的了解，分析中国海洋事业发展中各种失谐产生的深层次原因，并提出今后一个时期海洋管理转型的分析框架，同时对加快转型的机制进行初步设计。

一、和谐社会理论进展

"和谐"的基本含义是和睦、融洽、协调、均衡。作为一种状态和境界，它是处于矛盾漩涡中的人类与生俱来的理想，古今中外呼声不绝。但和谐的内涵、外延则与社会发展的诸多因素以及人们见仁见智有关。有斯宾诺莎以人权保障为核心的和谐，霍布斯、洛克以财产权安全为追求的和谐，孟德斯鸠以平等为理想的和谐，边沁以现实功利为指标的和谐，密尔的政府—社会良性互动式的和谐等。陶渊明的《桃花源记》，莫尔的《乌托邦》，康帕内拉的《太阳城》，[1-8]康有为的《大同书》，则在空想的基础上描绘了和谐社会的美丽画卷。

当代西方经济学以人的理性自利假说为圭臬，但没有淹没对和谐的追求。凯恩斯、罗宾逊、布兰查德、穆勒、伯格森和萨缪尔森、阿特金森等的社会福利、社会保障理论[9-12]旨在维护资本主义统治的前提下缓和社会冲突；罗马俱乐部（1987）在《我们共同的未来》报告中提出"可持续发展"，涉及人类代际和谐；日本政府的建设"环境友好型社会"，体现了人与自然和谐的理念。

中共十六届四中全会建立的"社会主义和谐社会"概念，标志着中国已经彻底地从阶级斗争时代转入了和平发展时代，实现了发展主义向价值理性的回归。近年有关研究与日俱增。

（1）研究了社会主义和谐社会的特征。胡锦涛提出六大基本特征（民主法治、公平正义、诚信友爱、充满活力、安定有序、人与自然和谐相处）；[13]成思危从哲学上概括为"三性"（统一性，大同小异；包容性，求同存异；调适性，增同减异），[14]还有的归纳为社会发展的整体性、社会结构的合理性、社会控制的有效性和社会关系的融合性[15]。

（2）探讨了实现和谐社会的路径。如陆学艺、邓伟志、胡位钧、蒋京议、高健、肖海鹏等提出了社会结构调整、利益格局均衡、公共服务均等；[16-21]尤其是探讨了顺应和谐诉求的管理改革，如胡锦涛（2005）提出"建立健全党委领导、政府负责、社会协同、公众参与的社会管理格局"；乌家培、李军鹏、郑应隆、徐永祥等提出了管理转型、治理模式创新等[22-25]。

(3) 沿着精确化方向量化了和谐社会的标准。如张德存给出了和谐社会的二层次指标体系和计算方法；[26] 牛文元给出了恩格尔系数（表征社会财富）、基尼系数（表征社会公平）、人文发展指数（表征经济社会协调）、二元结构系数（表征城乡一体化）、生产集约化弹性系数（表征资源增长速率）等指标[27]。

此外，向波强调构建和谐社会"是一个从不和谐到和谐，从局部和谐到整体和谐，从原有的和谐上升到新的和谐的动态平衡过程"。[28] 同时，以社会子系统为决策单元的相关研究（和谐广东、和谐社区、和谐校园、和谐交通等）日益增多。

二、理论观察下中国海洋领域中的问题

（一）现象描述

中国海洋经济总体水平已经跃升到世界海洋国家的中间偏上水平。海洋渔业、海盐产量连续多年保持世界第一；2007 年全国海洋生产总值 24 929 亿元，港口吞吐量世界第一；造船产量世界第二；中国沿海国际旅游人数、商船拥有量居世界第五位。[29] 然而，用上述理论观察，不难看到由于出现了"群龙闹海"的局面，海域利用的密度和强度大幅度增加，中国海洋事务中的失谐有某种上升趋势。王曙光主编的《蓝色国土忧思录》对此有多维度的描述。[30] 主要表现在：

（1）产业结构有失调。2005 年，海洋渔业、海洋交通运输业和海洋盐业这三大传统产业的产值占整个海洋产业总产值的 44%，海洋油气、化工、船舶三大工业相对薄弱，合计 8.2%。新兴的海洋制药、海水综合利用比重不足 10%。各产业内部的行业结构也不合理，如水产业的产品加工比例不到总产量的 40%，比照挪威等国家注重创造高附加值的产业形态，显然是效率损失。

（2）区域发展不均衡。广东、上海、山东 3 个地区主要海洋产业总产值之和占全国同比的 51%，其他 8 个沿海地区只占 49%。中国有海岛 6 000 多个，其中有人居住的只有 455 个。即便在相对发达的地区内，也存在"被发达遗忘的角落"。

（3）资源生态遭破坏。近四十年来，因大规模围垦丧失海滨滩涂湿地约 219 万公顷，相当于沿海湿地面积的 50%。20 世纪 50 年代中国有红树林约 5 万公顷，现只剩下约 2 万公顷。海南岛原来约 1/4 的岸段有珊瑚岸礁，现 80%

遭到了不同程度的破坏。2006年中国海域未达到清洁水质标准的面积为14.9万平方公里；全年共发生风暴潮、赤潮、海浪、溢油等海洋灾害179次，造成直接经济损失218.45亿元，死亡（含失踪）492人。[31]

（4）社区秩序未清平。用海中矛盾和纠纷很多。沙滩船厂、无证船舶取缔不尽；乱挖沙、乱倾废、乱设定置网、鱼栅、毒鱼、炸鱼时有发生；少数人信仰伦理已经被解体，海上盗、嫖、赌、毒、走私、黑社会等底层社会现象沉渣泛起。

（二）根源分析

1. 海洋系统的脆弱性

（1）海洋生物、矿产、化学、能源等资源的多元性和复合性，使得相关产业密集交叉。

（2）海洋生产力除了海岸、海岛、近海、远海的水平布局外，还有水面、水中、水下的垂直布局，造成多向度摩擦。

（3）海洋的"公共池塘"性，海上产权边界的模糊性，渔业资源的流动性，为零和博弈、负和博弈提供了诱因。

（4）同质而连续的水介质成为负外部性（如污染）传递的通畅管道。

（5）历史不长的海洋专门管理，与传统大陆型管理存在脱节，运筹中"海陆两分"。

2. 社会转型中的摩擦

社会转型伴随着利益洗牌，"破"与"立"很难平滑接榫。许多国家和地区的发展历程显示，人均GDP在1 000~3 000美元时期属于多事之秋。一些拉美国家人均GDP超过3 000美元，城市化率达70%，但由于贫富差距过大，导致经济社会震荡，被称为"拉美陷阱"。经济发展与公众满意度错位的"伊斯特林（Easterlin）悖论"[32]是其理论解释之一。

从实证分析，中国2007年人均GDP已经达2 666美元，沿海地区更高些，如广东人均GDP已经达到4 637美元。此时，经济体制深刻变革，社会结构深刻变动，利益格局深刻调整，思想观念深刻变化，全国经济社会呈现了"黄金发展期"与"矛盾凸显期"的交集。

经济力的增强具有双向后果：一方面，加大了社会矛盾的激发力。一是生产向广度的进军，囊括了更多的系统因子，而因子的增多必然使碰撞概率增加；二是根据马斯洛的需求层次原理，有支付能力的需求，特别是收入弹性大的需求爆发，造成了需求阶梯攀升中的拥挤效应；三是以加速度生成的新职业、新技术对劳动力的排挤，使结构性、摩擦性失业成为常态。另一方面，增强了解决社会矛盾的能量。如提高了人的素质、智慧、物质手段等。失谐是两种力量较量的表现。"拉美陷阱"并非铁律，而是由于矛盾协调的手段供给滞后于经济增长形成的特例。在海洋领域，主要是以下问题没有完全处理好：

（1）海洋开发目标片面。在大力开发海洋，建设"海上辽宁"、"海上山东"、"海上中国"的口号激励下，片面追求海洋经济总量的增长和外延扩大型增长模式，带来了多方面的失衡、脱节、冲突。"金色GDP"被"黑色GDP"切割。

（2）利益差别平抑乏力。中国沿海地区与历史形成的二元结构渗透在经济社会的方方面面，集中地表现为城乡差距。国家为了解决这个矛盾采取了许多改革措施，但是作为其根基的身份壁垒没有消除，不平等问题不可能真正消弭。大批农民工从内地流向沿海，为海洋经济发展节约了劳动力成本，但引发的新问题是，显在的"城乡二元结构"未解决，潜在的"城市二元结构"，即农民工与城市市民之间的地位鸿沟产生了。[33]根据笔者的观察，与此同时还存在一定程度的"农村二元结构"，即沿海养殖户雇用内地农民的情况。这三个"二元结构"，使规模性冲突具备了社会能量。

（3）公共池塘制度缺失。海洋自然资源具有准公共物品属性，被称为"公共池塘资源"（CPR），其难以分拨、流动性大等特性，使得其使用者存在规避责任、"搭便车"等机会主义行为，拥挤、退化、破坏、国有资产流失成为常态，这就是G.哈丁所说的"公地悲剧"[34]。中国虽然出台了渔业捕捞总量控制制度、养殖许可证制度、海域有偿使用制度，但是"土地"定价偏低，对利益相关者补偿制度不健全，操作中交易成本很高，委托—代理问题和道德风险存在。在笔者担任国家海域使用论证专家过程中，发现围海申请者多是具有某种背景的公司，有的是地方政府。他们实际是在做地产，而且"先斩后奏"，已经围海了才开始办证。

三、针对问题的海洋领域和谐社会建设目标

针对上述问题，2006年海洋行政主管部门提出了"和谐海洋"的口号，也有专家提出"和谐海洋社会"，本文不拟使用这两个词语，但赞同其主观上要表达的精神。认为：新时期海洋工作的基本目标，是追求以社会心理和谐为核心，经济关系和谐为基础，社会群体和谐和人与自然和谐为突出表现的海洋事业理想状态（见图1）。

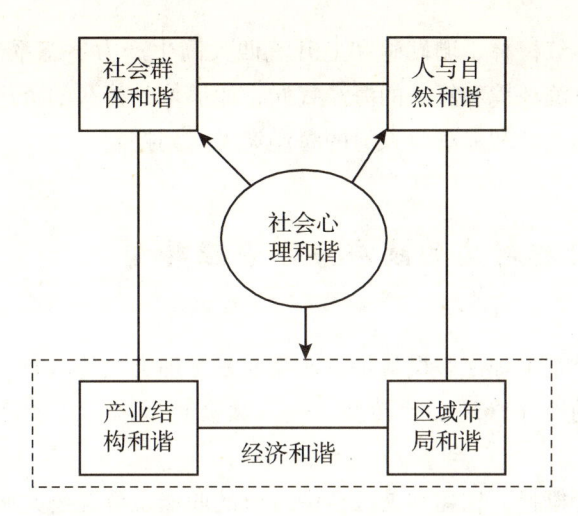

图1 和谐社会（海洋领域和谐）景观模型

（1）社会心理和谐。主要指人们的世界观、人生观、价值观、荣辱观趋同或相容。在海洋领域，是海洋价值观、海洋国土观在科学界、决策层和国民中都得到普及；海洋科学技术和海洋教育事业兴旺发达，海洋科技力量得到整合，形成充满活力的科技创新体系；海洋文化繁荣。先进文化成为海洋开发的精神资源、方向保证和是非标准。在观念层面形成"亲海、知海、用海、护海"的氛围，使海洋文化或海洋文明成为中华文明的基本组成部分。

（2）产业结构和谐。国家行业标准《海洋经济统计分类与代码》（HY/TO52-1999），把海洋产业划分为15个大类、54个中类、107个小类。这些产业之间的比例动态地保持匹配性，社会再生产得以顺利进行，既不因"短板效应"拖整个经济的后腿，也不因"过剩效应"造成资源浪费。

（3）区域布局和谐。海岸带、近海、岛屿、远海、专属经济区和大陆架经济都因地制宜地得到发展。沿海各省、市、自治区能够准确地找到自己在全局中的功能定位，通过主导产业发挥资源的比较优势，在与外部交流中实现互利、互补和发展水平趋近。

（4）社会群体和谐。海洋社会劳动者中的工、农、商、学、干，其劳动付出与物质、精神的回报虽有差别，但可以相互容忍；海洋法律、规章健全，居民行为自觉、规范，政府协调手段有力，安全标志和设施完备，突发事件应急体制健全，社会秩序井然。

（5）人与自然和谐。对海洋的馈赠心存感激，做到善待海洋、人海友好。摒弃无度索取、肆意掠夺行为。海洋资源的再生能力和海水的自净能力永续利用。陆源污染和海洋产业自身污染的势头得到遏制，海洋生物多样性得到延续，滨海湿地、红树林、珊瑚礁和上升流四大高生产力生态系受到重点保护，清洁、优美的海滨环境和舒适的海洋气候，能够作为持久的资产世代传递。地面沉降、海平面上升和赤潮等人为的海洋灾害得到控制。

四、与目标对应的政府海洋管理转型

本文的一个基本理论假设就是和谐绩效是管理类型的函数。为了实现和谐的目标，传统的海洋管理必须转型，并与社会转型耦合。这就是古人所说的"政通人和"。

据汪永成等概括，构建和谐社会中政府的职能定位是公共服务职能（包括制度供给职能）、社会管理职能和社会整合功能。[35]李荣娟把和谐社会构建中的政府功能归结为：保护、引导、教育、示范、协调和惩戒。[36]向波强调政府在构建和谐社会中的着力点是推进社会公平、整合社会关系和解决社会矛盾。[37]政府功能转变体现为管理转型。盖伊·彼得斯梳理归纳出四种未来政府治理模式：市场式（强调政府管理市场化）、参与式（主张对政府管理有更多的参与）、弹性化（认为政府需要更多的灵活性）、解制型（提出减少政府内部规则）。[38]中国学者也就管理转型问题提出了很多意见。

结合海洋领域的国情，笔者认为，中国政府海洋管理最重要的是实行"行政型→治理型、管制型→服务型、权力型→责任型、人治型→法治型、全能型→有限型、经验型→学习型"六大转变，它们与和谐目标之间存在着因果性的内在联系（见表1）。

表1　　　　　　　　管理转型与和谐绩效的对应关系

管理转型	和谐绩效
行政型→治理型	通过管理主体的多元化实现与管理对象的一致化
管制型→服务型	实现人民主人翁地位的回归
权力型→责任型	公仆职责得到保障
人治型→法治型	减少随意性和道德风险，增强和谐秩序的可靠性
全能型→有限型	收敛公民的被干涉领域，扩大其自治、自律空间
经验型→学习型	动态地打造和谐建设的知识支持

所谓治理型，就是以民为本，实行分权制度，政府、公共组织、私人机构及社会个人协同承担公共事务。管理方式多元化，划分政府、市场、社区在海洋开发中的功能边界，发育"公域"、"私域"之外的"第三域"——社会中

介组织、志愿者队伍，推动民间基层自治。

所谓服务型，就是勤勉、廉洁地为公众和全社会提供优质公共产品和服务。做"精明的引航员，公正的评判员，忠实的服务员"。要从海洋自然资源的不可分拨性出发，分析不同的制度安排下，人们不同的行为方式及其后果，从而选择海洋事业可持续发展的组织形式和政策。重点是以产权界定为核心，规范和改进伏季休渔、苗种放流、许可证、海域使用论证制度，建立海洋资源定价、配额转让、排污权转让、绿色GDP统计等制度。对渔村出现的"公司＋协会＋农户"、"科技入户"、"船东互保"等制度加以完善、推广。

所谓责任型，就是对人民高度负责，有效履行职责，对失职或行政不当承担相应的责任，损害当事人合法权益应承担赔偿责任。落实"过错责任追究制度"和引咎辞职制度。针对海上流动人口的活动特点，采取相应方式，为他们接受教育、医疗保健提供方便。防止所谓"蛋民"和鱼排人家的孩子辍学。加强海岛尤其是偏远小岛的交通、电力、学校建设，解决淡水供应问题，推行"造血式"扶贫。对因国际渔场划界、产业调整、征地等原因"失海"的渔民，从制度层面解读其权利，[39]提供再就业岗位、技术培训、法律救济。

所谓法治型，就是坚持行政权力只源自法律授权的行政法治原则，政府依照法律行使法律授予的职权，政府的行政行为，"法无规定不得为之，法定事项依法为之"。与国际海洋法律制度接轨，完善海洋法律体系，建设统一高效的海上执法队伍。

所谓有限型，就是建立科学的权力边界，政府收敛行动范围，有所为有所不为。集中精力管好法律规定的管理事项，凡是管理相对人能够自主决定、市场竞争机制能够有效调节、行业组织或中介机构能够自律管理的事务，政府都要退场。

所谓学习型，就是在内部形成浓郁的学习氛围，完善终身教育体系和机制，形成全员学习、团队学习局面，系统掌握海洋科学知识，提高专业水平，在学习中创新，增强群体工作能力。

五、推进海洋管理转型的基本机制

管理转型受到传统观念、习惯势力和既得利益集团的阻滞，有可能进展缓慢，甚至流于空谈。因此，必须建立强有力的推进机制。

所谓"机制"，笔者给出的命题式解释是：机制是一种装置的力量，能使宿主系统自动按照设定的模式运动。在社会生活中，此装置主要由一组规则和相应的物质手段构成。加快海洋管理转型的推进机制主要有利益激励拉力和制度约束、公众监督压力（见图2）。

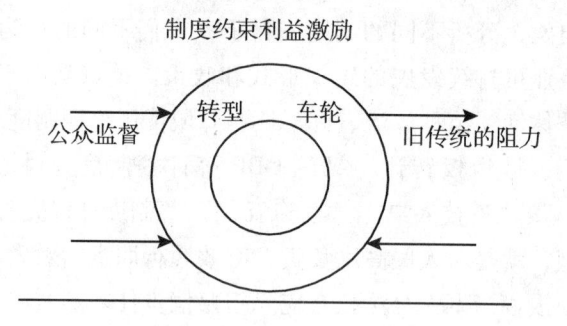

图2　海洋管理转型的受力模型

1. 利益激励机制

管理转型本质上是一种制度变迁。制度变迁的动力是获取体制外利润。丹尼尔·W. 布罗姆利把制度安排活动的动机分为四种：直接增加货币化净所得，重新分配收入，重新配置经济机会，重新分配经济优势。[40]总之，动力来源于对预设报偿的追求。

激励机制即诱导因素集合，是用于调动政府自身提高积极性的各种奖酬资源。闫平义提出，激励机制应该是人作为一种群体性、高智慧动物社会属性的表达方式。"心理高峰体验"是人类一种近于生理性的需求。随着社会的进步，高峰体验的价值系统开始多起来，除了物质财富、官职，还有国家级奖、名人地位、公众赞誉等。求"德"（贡献大于报酬）也是一种高峰体验。[41]据此，应该把道德自豪或自慰纳入管理转型诱导集合。如从海洋领域的实际和特点出发，加强对管理者思想、道德的教育培养；干部的薪酬、晋升，更多地与管理改革的绩效挂钩；营造"改革光荣、奉献伟大"的舆论氛围；对创造管理新模式的，给予宣扬，甚至树碑立传。

2. 制度约束机制

制度或游戏规则，包括正式、非正式两种，发挥着法治、德治两种功能。制度的价值，是通过个人收益率与社会收益率的统一，鼓励"生产性努力"，约束"分配性努力"。[42]

以上管理的转型，凡是中央或社会公认的，都应该通过立法、章程、公约等形成正式规则，明确标准和运行程序。例如，各级海洋管理机构设置、权限、职责、办事程序等改革，都要有制度化的表达，有条件的还要建立相关评介指标体系。

3. 公众监督机制

人民群众与政府管理机关之间本来就是委托—代理关系，管理行为的后果直接影响人民的利益，因此天然存在着"监督饥渴"。但是由于强制力不在公众一方而在政府一方，公众反而居于弱势，丧失话语权。必须让权力在阳光下运行。例如对养殖滩涂的承包、航道的建设、海域的占用、小岛的整体开发等，要实行政务公开、行政公开、执法透明；建立和完善公共信息披露制度、听证制度、法定权利告知制度、行政诉讼制度、电子招标制度；抑制分利集团的偏好；结合发育第三方独立机构，有组织地开展评议政府工作绩效、改革绩效的活动。

参考文献

[1] 斯宾诺莎. 笛卡尔哲学原理 [M]. 王荫庭，洪汉鼎，译. 北京：商务印书馆，1991.

[2] 霍布斯. 利维坦 [M]. 黎思复，黎廷弼译，杨昌裕，校. 北京：商务印书馆，1999.

[3] 洛克. 政府论 [M]. 叶启芳，瞿菊农，译. 北京：商务印书馆，1981.

[4] 边沁. 政府片论 [M]. 沈叔平，等译. 北京：商务印书馆，1996.

[5] 孟德斯鸠. 论法的精神 [M]. 张雁深，译. 北京：商务印书馆，1982.

[6] 托马斯·莫尔. 乌托邦 [M]. 戴镏龄，译. 北京：商务印书馆，1982.

[7] 康帕内拉. 太阳城 [M]. 陈大维，黎思复，黎廷弼，译. 北京：商务印书馆，1982.

[8] 约翰·密尔. 论自由 [M]. 程崇华，译. 北京：商务印书馆，1982.

[9] 琼·罗宾逊. 现代经济学导论 [M]. 陈彪如，译. 北京：商务印书馆，1983.

[10] 约翰·穆勒. 政治经济学原理 [M]. 赵荣潜，等译. 北京：商务印书馆，1991.

[11] 萨缪尔森. 经济学 [M]. 高鸿业，译. 北京：商务印书馆，1983.

[12] 阿特金森. 公共经济学 [M]. 上海人民出版社，1996.

[13] 胡锦涛. 在中共中央举办的省部级主要领导干部提高构建社会主义和谐社会能力专题研讨班开班式上的讲话 [J]. 党建，2005 (3-4)：1-3.

[14] 成思危. 和谐社会理念的哲学基础 (N). 人民日报，2005-9-29.

[15] [35] 汪永成，靳江好，李继琼. 构建和谐社会进程中政府的职能定位与能力建设 [J]. 行政管理，2005 (10)：79-82.

[16] 陆学艺. 构建和谐社会与社会结构调整 [J]. 江苏社会科学，2005 (6)：

9-11.

[17] 邓伟志. 如何构建一个和谐社会——在浙江省委党校的讲演 [N]. 文汇报, 2005-1-11.

[18] 胡位钧. 论当前我国社会结构的分化与协调 [J]. 改革与发展, 2005 (6): 80-84.

[19] 蒋京议. 调节社会利益关系的制度机制 [N]. 中国经济时报, 2004-04-29.

[20] 高健. 利益均衡——转型期社会和谐的必要条件 [J]. 探索, 2005 (5): 132-135.

[21] 肖海鹏. 积极推进基本公共服务均等化 [N]. 南方日报, 2006-12-21.

[22] 乌家培. 管理转型与转型管理 [J]. 学术研究, 2006 (5): 5-7.

[23] 李军鹏. 和谐社会建设与社会治理模式创新 [J]. 国家行政学院学报, 2005 (4): 82-84.

[24] 蔡茂生, 郑应隆. 管理转型: 因应发展新阶段的选择 [N]. 南方日报, 2006-04-13.

[25] 徐永祥. 社会体制改革与和谐社会构建 [J]. 学习与探索, 2006 (6): 18-21.

[26] 张德存. 和谐社会评价指标体系的构建 [J]. 统计与决策, 2005 (11).

[27] 和谐社会的五大基本标志 [N]. 中国劳动保障报, 2006-3-15.

[28] [37] 向波. 和谐社会视域中的我国社会治理创新 [J]. 探索, 2005 (6): 93-96.

[29] 国家海洋局. 2007年中国海洋经济统计公报 [EB/OL]. 2008-04-11. http://www.soa.gov.cn/hygb/2006jingji/7.htm.

[30] 王曙光. 蓝色国土忧思录 [C]. 北京: 海洋出版社, 1999: 5, 15, 18, 21, 41, 45, 48.

[31] [32] 国家海洋局. 2006年中国海洋环境质量公报 [EB/OL]. 中国海洋信息网, 2007-5-9.

[33] 田国强, 杨立岩. 对"幸福——收入之谜"的一个解答 [J]. 经济研究, 2006 (11): 4-15.

[34] 邓伟志. 如何构建一个和谐社会——在浙江省委党校的讲演 [N]. 文汇报, 2005-1-11.

[35] Garrett Hardin. The Tragedy of the commons [J]. Science, 1968, 162: 1243-1248.

[36] 李荣娟. 和谐社会构建中的政府功能 [J]. 当代世界社会主义, 2005 (6):

77-81.
[37] B. 盖伊·彼得斯. 政府未来的治理模式 [M]. 北京：中国人民大学出版社，2000.
[38] 全永波."失海渔民的权益缺失与法律救济 [J]. 海洋开发与管理，2007（5）：47-52.
[39] 丹尼尔·W. 布罗姆利. 经济利益与经济制度 [M]. 陈郁，等译. 上海：上海三联书店，上海人民出版社，1996.
[40] 闫平义. 社会发展动力机制的科学设计 [EB/OL]. 2005-01-09. http://publishblog.blogchina.com/blog.
[41] 萨缪尔森·B.，诺德豪斯·W. 经济学 [M]. 萧琛，等译，华夏出版社，麦格劳-希尔出版公司，1999.

Transition of Government's Role in Ocean Administration in the Collective Effort to Build a Harmonious Society

Xu Zhibin

[Abstract] In the light of building a harmonious society, we can observe the "inharmonious" spots in ocean management. These standing issues can be attributed to the vulnerability of the complex system of ocean administration, and more importantly, to problems caused by inadequate administration of a society in transition. The primary goal of ocean administration should be centered on the pursuit of harmony in social psychology and based on a harmonious economic relationship. Success in this aspect is typically demonstrated in harmonious relationships between various social groups and that between Human and Nature. "Harmony index" belongs to the sphere of administration. The government's role in ocean administration should respond to the public appeal for harmony, and therefore should try to accomplish the transition in several dimensions: from simple and mandatory administration to service-oriented administration and management, from power to responsibility, from man rule to law rule, from all encompassing administration to focused administration, from being empiricist to experimentalist. In addition, incentive mechanism, relevant regulations and public supervision should be in place to advance the cause.

[Key Words] Harmonious society　Ocean administration　Transition　Mechanism

JEL Classification: A14, E69

中国海洋经济理论演化研究

姜旭朝　黄　聪[*]

【摘要】 本文从海洋经济学自身的理论本质入手，对新中国成立以来各个时期的海洋经济学理论研究发展的总体规律以及各个时期的不同特征进行了探讨，分析了其定义、方法论和理论体系结构的发展脉络，总结了其演化的一般规律，即点—线—面—空间的系统性发展演化进程，并对其未来发展方向作了简要展望。

【关键词】 海洋经济学　经济思想史　演化

一、序言

从历史角度讲，人类海洋经济活动的历史并不短，但从经济的角度上研究这些活动则并不长。直到20世纪40年代，美国和日本才出现了一些对于海运经济和渔业经济的零星研究。第二次世界大战后，部分发达国家才开始较为重视海洋经济管理的研究。但所有这些研究都是局部性的和单项性的，没有把各种海洋经济活动作为一个整体，从中考察它们的联系和运动。理论的产生与发展是与实践紧密相连的，而海洋经济理论的产生和演化与海洋经济体系的产生、发展、成熟有着十分紧密的联系。20世纪60年代，随着人类社会经济体

---

[*] 姜旭朝，中国海洋大学经济学院院长、教授，山东青岛：200071，Email: jiangxuzhao@ouc.edu.cn；黄聪，中国海洋大学经济学院2006级研究生，Email: huangcong1113@yahoo.com.cn。

系不断扩张，陆上自然资源开发利用已很难满足需求，各国开始将着眼点放到广阔的海洋之上，构成海洋经济产生的必要性；另外，科学技术水准的发展，特别是海洋相关科技的发展，使人类大规模开发、利用海洋资源和空间具备了物质基础，这是海洋经济产生的可能性。在此时期，海洋产业种类不断增多，规模不断扩大，在经济体系中的地位也日益重要。海洋相关各产业已经不单单是陆地经济体系的一部分或者补充，而是开始构成一种独立的、可以与陆地经济分庭抗礼的海洋经济体系。在这个过程中，海洋经济不同于陆地经济的特点逐渐显现，传统的、以往建立在陆地经济体系下发展起来的经济学理论体系对海洋开发实践中所面临的各种问题解释力也就愈发不足，这就要求对其进行改良甚至变革，从而最终产生了海洋经济学理论。

传统的经济理论虽然是人们对经济现象、经济行为本质规律的研究，与海洋、陆地的差别并无关系。但不可否认，这套体系是在陆上经济占绝对主导地位时期产生、完善起来的。由于此时人类对经济活动认识的片面性，其不可避免地存在着各种与陆地经济特点相联系的片面观点、理论，我们可以将其称为经济理论的"陆上痕迹"。而综观我国海洋经济研究历史的变迁情况，可以发现无论是从对海洋经济、海洋经济学本身认识与理解的发展过程之中，还是在整个海洋经济理论研究体系的扩张进程之内，都体现出一种新的研究领域、框架、范式的孕育、发展和成熟的过程，其发展趋势必然是建立一个与传统陆域经济研究体系截然不同、可以分庭抗礼的海洋经济体系。但存在这种趋势并不是说海洋经济应该完全脱离现有的经济分析体系独立发展，海洋经济理论的发展演进，归根结底仍然是对原有经济理论的运用、检验、批判和改革的过程。在这个过程中寻找、辨识和消除其中的"陆上痕迹"，建立起一套更适应现代经济要求的基本经济理论体系，才是海洋经济研究最为本质的目的。

本文从海洋经济学自身的特征入手，对新中国建立以来海洋经济学理论研究的总体规律、发展脉络以及各个时期的不同特征进行了探讨，总结了理论演化的一般规律，并对其外来发展方向作了简要展望。全文共分六个部分，第一部分为序言，对海洋经济学产生的社会背景及其理论本质进行论述；第二部分介绍了中国海洋经济理论演化的总体趋势，包括对其定义、理论体系和方法论演化进程的研究；第三部分分析了在我国系统海洋经济学产生之前，即所谓海洋经济学"萌芽时期"产程的原因及其特点；第四到第六部分，则是将我国海洋经济学自提出30年以来的研究历程大致分为三个阶段，对各时期特点与趋势进行了系统的分析，并对海洋经济研究的未来发展趋势作了简要展望。

二、中国海洋经济理论演化的总体趋势

海洋经济的发展壮大与其不同于陆地经济学的特点，促使了海洋经济学的产生。海洋经济学是随着海洋经济活动的不断深入而发展起来的一门学科，人们对其研究范畴的认识也是在不断发展演进着的。"海洋经济不是简单的陆地经济向海洋的延伸和重复，而是一个全新的经济领域，它有着与陆地经济完全不同的特点与运行规律。这样一来，即使时间以经济提出了这方面的需要，一个整体的和系统的海洋经济理论体系也不可能立即形成。它需要全面总结国内外的海洋经济实践经验，并上升到理性认识，这项工作绝非少数人一朝一夕就能够办到的。"而我国海洋经济学理论30年的发展历程，恰恰就经历了一条点—线—面—空间系统化的理论发展演化脉络。

总体来说，我国海洋经济研究的历史可以分为前、后两大阶段，其分界线为1978年。1978年在全国哲学社会科学规划会议上，一些学者提出建立"海洋经济"学科和专门研究机构。1949~1978年，虽然我国并未提出"海洋经济"这一概念，但此时国内已经存在着规模不小的海洋相关产业，如海洋渔业和海洋运输业，因此出现了一些围绕这些产业的个别问题进行研究的成果，但较为零散，文献散见于渔业经济、运输经济或农业经济各研究领域之中，并未形成系统的研究方法和学科体系，因此可以称此为我国海洋经济研究的"萌芽时期"。1978年著名的经济学家于光远、许涤新等在全国哲学和社会科学院规划会议上提出建立"海洋经济学"新学科以及针对其的专门研究所。1980年7月，我国召开了第一次海洋经济研讨会，并且成立了中国海洋经济研究会。

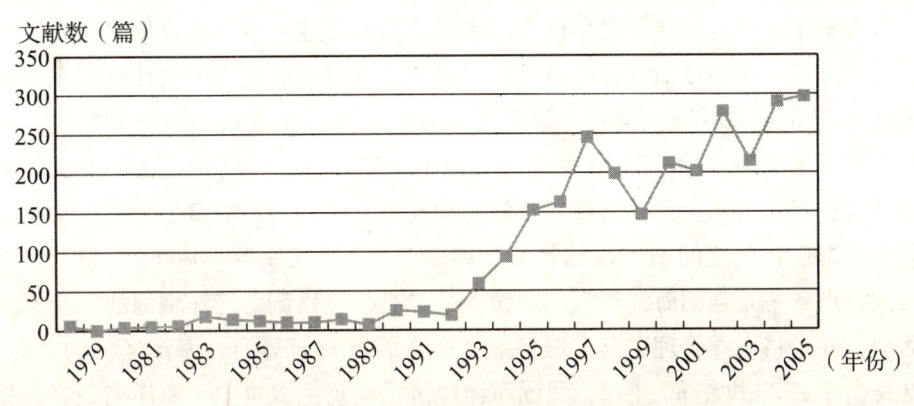

图1 1978~2007年海洋经济相关文献数量变动图

以此为标志，我国海洋经济体系逐步建立，相关研究工作开始踏上正轨，并涌现了一大批杰出的专家、学者。这种趋势仅从文献数量就可以看出（见图1），从1979年开始我国海洋经济相关文献数量不断上升，先后经历了20世纪80年代的平稳增长阶段、90年代的迅猛发展阶段以及21世纪之后的成熟阶段，至此我国海洋经济的学科体系已日臻完善。

（一）海洋经济范畴的演变

"海洋经济"范畴是海洋经济理论研究的基础。从公开发表的文章来看，我国学者对于海洋经济的定义有着明显的时代特征，是随着海洋开发实践和相关研究工作的不断深入而不断变化着的，而且自出现以来其内涵就在不断扩大。迄今为止，学术界对海洋经济的概念尚没有形成公认的一致性看法。从学者们的总体研究成果看，海洋经济的范围经历了一个由窄到宽、由资源性到产业化、由陆域经济体系的附庸到与其对立的新的经济体系的升级过程。

在我国，海洋经济这个概念最早是由著名经济学家于光远在1978年提出的，他在全国哲学和社会科学院规划会议上提出建立"海洋经济学"学科的建议，并建议建立一个专门的研究所。1980年7月，在著名经济学家中国社会科学院经济研究所所长许涤新亲自指导下，召开了我国第一次海洋经济研讨会，并且成立了中国海洋经济研究会。此时，"海洋经济"这个词才广泛出现在各种专业论文上，但此时还没有一个对其系统、完整的定义。直到1982年，才有人第一次提到海洋经济的定义。何宏权、程福祜（1982）的文章中提到："所有这些人类在海洋中及以海洋资源为对象的社会生产、交换、分配和消费活动，统称为海洋经济。海洋经济活动的范围是在海洋，就空间地理位置来说有别于陆地，故称海洋经济。"

在整个20世纪80年代，许多学者都从自身专业背景、研究思路及实践经验入手，从不同角度对海洋经济进行了定义，并且有着明显的时代性特征。在早期（20世纪80年代中期），学者主要立足于社会主义政治经济学角度，将海洋经济定义为："海洋经济活动是人们为了满足社会经济生活的需要，以海洋及其资源为劳动对象，通过一定的劳动投入而获取物质财富的劳动过程，亦即人与海洋自然之间所实现的物质变换的过程"（权锡鉴，1986）。该定义有两大特点：一是强调海洋经济的资源性；二是仅仅强调对海洋资源的一次开发活动，并未将海洋管理、海洋服务业、海洋教育科研等列入海洋经济的范畴。这是由于在20世纪80年代，我国海洋经济活动仍主要集中于海洋捕鱼、海盐制造、海洋运输等传统产业，而国家对于海洋开发的远景规划也仅限"海洋是一个巨大的资源宝库，存在着众多的矿产、石油资源"。如何合理高效地开发各种海洋资源也就成为了海洋经济学者所要研究的问题。

20世纪90年代之后，在经济领域中，海洋经济开发活动不断深入；而在思想领域中，各种西方经济学的思想、方法进入中国，极大地扩展了学者们的视野。致使学界对"海洋经济"的看法也呈现出百花齐放的态势。此时，多数学者仍然从资源经济的角度进行定义，从本质上讲，仍然是将海洋经济看做陆域经济的附属（资源产地），如陈万灵（1998）、周江（2000）、徐杏（2002）、陈学林（2007）等；也有部分学者坚持资源经济论的同时，注意到以海洋空间作为活动场所的经济行为规模不断上升的客观现实，并将其纳入了自身的理论中去，如杨金森（1984）、陈可文（2001，2003）、何翔舟（2002）、曹忠祥等（2005）；或者是从资源开发角度出发，从产业发展的角度对海洋经济进行定义，将其视为围绕海洋资源进行了一系列生产、分配、交换、消费活动的总和以及其所形成的一系列上下游产业，如许启望、张玉祥（1998）、陈可文（2003），而这也是我国官方所最为承认的观点，在许多正式的文件、标准中所常常使用的（《全国海洋经济发展规划纲要》，2003；《海洋学术语——海洋资源学》，2004）；这些观点初步摆脱了海洋经济作为陆域经济附属和延伸的地位，开始将其视为对陆地经济体系的重要补充和组成部分。

也有学者从海洋经济与陆地经济的对立入手，从区域角度来确定其范畴，因为广义上的海洋经济，主要是指与海洋经济难以分割的海岛上的陆域产业海岸带的陆域产业及河海体系中的内河经济等，包括海岛经济和沿海经济（陈可文，2003）。有人认为海洋经济实质上就是区域经济，如"海洋经济包括海岛经济"，甚至认为向陆地若干公里以内的海岸带经济均属于海洋经济的范畴。还有学者从沿海区域资源经济、产业经济和滨海区域经济相结合的角度来理解海洋经济的内涵，认为"从科学、系统的角度理解，它是对沿海区域资源经济、产业经济和滨海区域经济的有机综合；发展海洋经济是以海洋资源为基础，以海洋产业为桥梁，以沿海区域的社会经济全面发展为目标的一项系统工程"（郑贵斌，2006）；或者从最广义的角度将海洋经济定义为，"是产品的投入与产出、需求和供给与海洋资源、海洋空间、海洋环境条件直接或间接相关的经济活动的总称"（徐质斌，1995）。此外持类似观点的还包括：孙斌、徐质斌（2000），徐质斌、牛福增（2003）等。这些观点将海洋经济在体系上从陆地经济中彻底剥离出来，成为一个与其同等地位的经济体系，也从客观上说明了对海洋经济这一特殊经济体系相关理论研究的必要。

（二）海洋经济理论体系发展趋势

　　早期的海洋经济理论研究多具有离散的特点，可以分为产业（行业）研究、区域研究（此时海洋区域研究是指"小区域"研究，往往研究的是一省、一市甚至一县的海洋经济问题）资源开发研究，缺乏思维张力，忽视多维空间和整体设计，但其整体上与当时海洋开发水平是协调的。研究区域仅仅限定在海岸线的界定区间或者近海、浅海区域，从区域上并没有和陆地分开。名为研究海洋经济，实为研究陆地附近经济或浅海经济。从经济研究的半径与成本来考虑，在海洋经济欠发达或不发达时期，由于其专门研究所能带来的社会效益并不显著，为降低政府在进行相关研究活动所投入的直接和间接成本，将海洋经济理论界定于各部门经济研究之中是值得的。然而，当社会经济进一步发展，科学技术成果以几何级数增加，海洋科学与海洋产业向纵深化、全方位拓展并成为与大陆经济平行存在与发展的趋势下（它虽然在产出量上与大陆经济还无法比较，但从资源存量及开发潜力上比较，其前景比大陆经济宽广得多），再将其归属为大农业经济或部门经济就只会制约其发展前景及研究水平的提高。因此，随着我国海洋开发实践的不断进步，海洋经济研究进程也沿着纵横两条轨迹不断扩展和演进。

　　横的轨迹是指，对海洋经济理论的认识是通过对其辖下单个部门经济学不断综合的演进过程。在海洋经济发展的初级阶段，海洋是作为一个综合性的资源载体而存在，其资源的开发是分门别类逐渐进行的，而且是由各个行业、各个部门分别从陆地向海洋延伸而进行的，由此形成海洋经济活动从一开始就是分散地进行。因此，海洋经济研究必然会首先产生于各个行业或部门经济学之中。在20世纪50~60年代，海洋经济理论仍未成型，此时的海洋经济研究成果散见于农业经济（海洋渔业经济）、运输经济（海洋运输）、工业经济（海洋盐业）等其他相关经济科学的研究成果中十分零散。1981年，随着整个海洋经济学科的建立，许多学者就将这些零散的研究成果简单加合起来，形成了最初的海洋经济理论，并将其看做一种综合性的部门经济学。因此，此时的各种研究成果往往为对个别行业的个别性研究，即使综合性文章往往也仅仅是将不同方面的研究简单组合起来。对海洋经济的认识狭小，无疑将限制了人们研究海洋经济的视野。由于在研究思路方面拓展不开，必将影响到研究层次的深浅与研究成果的多寡。

　　到了20世纪80年代后期，有部分文献开始注意到进行综合性研究的重要性，海洋经济学的研究对象，就是从宏观上研究如何合理地、有效地开发利用海洋生产力运动规律的科学，在全局利益和长远利益的前提下，全面协调、相互配合、合理地利用海洋资源，有效地组织生产，才能提高经济效益。到了20世纪90年代以后，许多学者开始利用西方宏观经济、产业经济的相关理论，将涉海各经济部门进行横向综合研究，将海洋经济理论研究扩展到包括关于海洋经济各方面研究工作的一切科学成果。包括："关于海洋经济整体及各部门和区域的本质、特征、结构、规律的研究成果；关于海洋经济各侧面（生产力、生产关系、管理体制等）的研究成果；关于海洋经济各运行环节的研究成果；关于海洋经济条件的研究成果；关于海洋经济发展历史的研究成果；关于海洋技术经济的研究成果；关于海洋工程经济的研究成果等"（徐质斌，1995）。至此，海洋经济学已不只是关于海洋渔业、海洋运输业、海洋资源开发业、海洋服务业和海洋旅游等个别海洋经济领域的理论，而成为以整个海洋经济系统的运行作为研究对象，以合理控制、科学协调和维护国家海洋经济整体利益为目标的经济理论。也就是说，海洋经济学的研究对象是海洋经济综合体或整体，其已经成为一门综合性科学。

而纵的轨迹则是从海洋资源的开发和利用出发，将经济科学与海洋科学结合起来，逐渐由海洋资源开发的研究向海洋资源的深加工，再到资源的产业化发展，资源产业化对区域经济、环境经济的影响，最后发展到对国民经济和世界一体化趋势的影响研究，最终从纵向贯穿海洋经济理论的整个体系。在早期，持此种观点的学者往往将海洋经济科学视为自然学科与社会科学，或者说是海洋自然科学与理论经济学的"边缘科学"，主要研究的是资源开发的经济效益问题。随着海洋经济实践的深入，海洋经济研究向纵向不断开始，一些学者开始将其视为"以经济学、政治经济学和生产力经济学为理论基础的"、"一门应用经济学"，"不能把海洋经济学划入边缘经济学种类"，"因为它是把理论经济学的基本原理应用于海洋资源开发与利用的实践，在实践的基础上进行经济总结、理论抽象与揭示客观规律，并为海洋资源的开发利用和保护服务的学科"（孙斌、徐质斌，2000）。进入21世纪，则更进一步将海洋经济科学定义为有关海洋开发与管理、利用与保护、改造与培育过程中特殊经济问题的一门学科。

总体来说，"纵"的轨迹是以纵向不同产业，不断通过水平运动向相关产业机制扩张进行理论发展；"横"向发展则是以产业链为基础，从海洋资源的开发出发，逐渐向其后的关联发展，从资源一次利用，再到深加工，进一步到相关服务、科研教育活动，都一步步地纳入海洋经济理论研究的范畴。但"纵"向发展与"横"向发展不是相互割裂的，其相互之间有着或多或少的"交叉点"，且随着两种轨迹的不断深入，这种"交叉"部分不断增加，从而促进两种理论的相互融合。

直到进入21世纪之后，随着海洋经济实践进入新的阶段，海洋经济理论横、纵两条发展主线已逐渐交织在一起，开始有学者将海洋产业经济观点与海洋资源经济观点融合起来，建立起海洋"大区域经济理论"（此种观点即将海洋与陆地对立起来，将其视为独立的一块"区域"进行研究）；也有学者运用最新发展起来的集成创新理论与系统创新观，按照社会经济活动的纵横结构规律和经济科学的分类要求，将海洋经济学定义为是在海洋空间范围内人类的经济活动与各种海洋因素之间相互关系的一门科学，将海洋资源环境经济、海洋产业经济和海洋区域经济结合为一个密切联系的系统，最典型的就是郑贵斌（2005）提出的"海洋经济位理论"。由于此时期，海洋经济是相对于陆地经济提出的概念，因此其内涵已是非常丰富的，也比较全面地反映了海洋经济本身的各种属性。但从经济学科发展的规律来看，海洋经济理论仍然仅限于以现有的理论体现来解释"涉海"经济活动，而未对经典经济理论本身有所补充或修正，仍然处于理论发展的初级阶段。

（三）海洋经济学研究方法论的演化

任何一门学科都有与其学科特点相适应的研究方法，方法的正确与否对其未来发展影响重大。海洋经济学作为一门新的学科，它的研究方法本身亦是重要的研究课题。由于理论研究工作实践并不长，还没有人专门探讨它的研究方法。因此，系统地回顾海洋经济研究的方法论，有助于进一步寻找和改善其研究方法。

海洋经济研究的方法论，基本经历了调查研究基础下的定性分析（1949~1990年）——基本经济理论指导之下的理论定性分析（1991~2000年）——具体经济学理论与先进的经济学模型为基础的定量分析（2001~至今）这三个步骤。体现了方法论由低级向高级跃迁的演化过程。虽然说在这种跃迁中明显受到了其他外来因素的干扰，如文化大革命，如西方经济理论的逐步引入，但从本质上讲，出现这种跃迁轨迹并不是偶然促成的，而是海洋经济特性所决定的。最为显著的特征就是，除了海洋经济宏观体系体现出了这种特征，在具体到个别海洋部门的经济理论研究中也体现出了这种特点，如在20世纪90年代才开始出现的海洋旅游经济研究的演化进程之中，就出现了这种演化进程，即首先对各地海洋旅游状况进行调查、分析，提出制约因素和解决方法（20世纪90年代前期）；其次，引入了西方旅游经济的一般理论对其作深入的理论分析（20世纪90年代后期）；最后，开始构造数量模型进行因素分析（21世纪之后）。可以说，海洋旅游经济研究十余年的演化过程，与海洋经济理论近五十年的发展过程，有着惊人的相似。因此，可以认为，海洋经济研究的这种特性是由其自身的综合性、国际性、科技性特征造成的。这与海洋经济这种特征形成的海洋经济活动具有十分复杂的影响因素，很难在以陆域经济为主的传统经济理论中找到现成的理论进行解释，只能在问题出现之时逐步将原有理论进行调整，才能进行研究。

三、1949～1978 年，中国海洋经济研究的"萌芽"时期

在1978 年之前，我国还没有出现"海洋经济学"这一名词。此时对海洋经济的研究都是附属于陆地经济的相关领域中，往往是作为部门经济的特殊形式（如农业经济中的海洋渔业经济、运输经济中的海运经济），或者管理经济学的特定对象（如管理经济中的海岸带管理）来研究。其研究成果也并未形成系统，仅仅是零散地现于各种相关著作中。此时期，可称之为海洋经济研究的"萌芽时期"。

（一）海洋经济"萌芽"时期产生的原因

马克思说："人类始终只提出自己能够解决的任务，因为只要仔细考察就可以发现，任务本身，只有在解决它的物质条件已经存在或至少在形成过程中的时候，才会产生。"在20 世纪80 年代之前，人们对于海洋经济的认识，受到社会经济实践的影响，仍仅仅局限于一些较为零散的领域，如海洋渔业经济、海洋运输经济、海岸带管理经济，缺乏系统地针对海洋经济特点和规律的研究体系。这说明，即使在20 世纪70 年代末，相关的物质条件还没有形成，就像物质生产的产生与发展要受到供需两方面条件的制约，海洋经济理论的建立同样取决于这两种因素。

从需求角度讲，海洋经济学理论产品的需求者可以大体分为三类：公众、政府和企业。

政府可以说一直是经济学思想产品的大主顾。政府的决策活动中有相当一部分是制定经济政策或与经济有关的其他政策。但在20 世纪70 年代以前，从总体上看还局限在小范围内，行业间的矛盾还不太突出，国家间的海洋权益斗争尚不尖锐，因此，对海洋经济的研究还没有引起大多数国家的注意。国家海洋观念总体上的薄弱，对一项涉及许多行业和部门具有极强的综合性经济理论

的建立是一个非常突出的障碍。没有统一或综合的需要,当然也就不可能有综合的海洋经济理论的建立。并且此时期,受到"大跃进"和文化大革命的影响,政府在海洋管理方面的职能受到极大影响,对其投资处于停滞状态。因此,对相应经济理论的需求也趋于停滞。

企业同样是思想产品的主要购买者,企业不仅需要经济学家提供相应的政策指导,同时也需要经济理论本身作为其进一步发展壮大的指导。在此时期,我国的绝大部分涉海企业均为国营企业(主要为国营渔业公司、国营盐场以及交通部下属的海洋运输企业),生产方面的软约束使之对于进一步增加经济效益并非十分热衷,同样很难产生对经济理论的指导需求。

公众之所以成为经济学思想产品的需求者,是因为他们不论以什么身份出现在经济活动中——消费者、要素所有者、厂商的管理者,作为经济活动的参与者和决策者,都需要能指导其行动的政策建议。但当具体到当时的背景之下,由于国家海洋产业发展薄弱,其经济效应在国民经济中所占的比例非常小,且往往依附于其他产业之中(如海洋渔业经济属于农业经济,海洋运输经济依附于交通运输经济),公众对海洋经济活动几乎没有任何接触的机会,自然不可能对其背后的经济规律产生需求。

在供给方面,海洋经济理论思想产品的供给同样取决于投入,这种投入应当包括具有专业素质、符合海洋经济研究要求的经济学家和相应的海洋知识资本积累。

经济学家是经济学思想得以生产的必要条件。海洋经济不是简单的陆地经济向海洋的延伸和重复,而是一个全新的经济领域,它有着与陆地经济完全不同的特点与运行规律。这样一来,即使实践已经提出了这方面的需要,一个整体的和系统的海洋经济理论体系也不可能立即形成。要想给提供这种"供给"的经济学家以特殊的要求,需要他们不仅对经济理论本身有所了解,并且要熟悉海洋经济活动的各种规律、特性,相对于其他经济门类而言,增大了其所需的投入。在20世纪80年代之前,受国家政治(如反右斗争扩大化、文化大革命等对经济工作者的冲击)、经济(由于国民经济水平的落后,无法给经济工作者提供合理的收益)以及思想(缺乏对系统经济理论与海洋科技理论的传播与流通机制)等因素的冲击,很难提供足够的经济学家。

除了经济学家以外,知识资本是决定经济学思想产品供给潜力的第二个因素。思想产品不能凭空制造,必须借助以往已经生产出来的思想产品来制造,就像物质产品的生产一样,正是在这种意义上,称以往生产的思想产品为知识资本。它也是思想产品得以生产的必要条件。海洋浩瀚而复杂,存在着大量的未知因素，难以开发，造成海洋经济长期发展缓慢，至今仍然是一项不成熟的经济；实践上的不成熟，也难以形成成熟的理论。而在此时期，海洋经济活动仍然局限在很小的范围内。海洋是一个综合性的资源载体，资源的开发是分门别类逐渐进行的，而且是由各个行业、各个部门分别从陆地向海洋延伸而进行的，由此造成海洋经济活动从一开始就是分散地进行，并未引起人们的关注，缺少对其系统、全面地调查、分析与研究，没有积累足够的思想产品作为依托，海洋经济很难发展起来。

（二）海洋经济研究"萌芽"时期的特点

在此时期，海洋经济研究成果的供需状况决定了系统的海洋经济学科体系在此时期不可能建立起来，只能呈现出一种"萌芽"的状态，这包括以下几种特点：

第一，研究成果数量稀少。1949～1979年的30年间，笔者所收集到的海洋经济相关分析为59篇，数量上仅相当于1994年一年的文献发表量，且体现出显著的时代性特征，即绝大部分研究成果集中于1949～1960年和1973～1978年两个时期，1960～1973年的13年间，海洋经济研究的成果仅有3篇。这种情况的出现主要是受反右倾活动和文化大革命等活动的影响，20世纪60年代开始我国各学术刊物先后停刊，大批学者受到冲击，无法进行正常的研究工作，更是对海洋经济研究的发展构成了致命的冲击。在这种背景下，许多研究往往就是简单地对国外海洋产业发展成果的介绍与总结，且基本都是源引自国外经济学期刊，主要是各杂志采编了一些对其他国家的海洋渔业、海洋航运业发展历程的分析性、评论性的文章，如对法国海运业、① 挪威海运业、② 朝鲜海洋渔业、③ 苏联远洋渔业④等国家/产业的论述。这些文章基本都集中在20世纪50年代后期和1973～1975年前后，恰恰是我国具有较强的海洋产业发展意愿的时期，这为我国海洋产业研究较为闭塞的时期，带来了些许新意。

第二，从研究对象看，较为集中宏观研究领域，而具体研究内容则较为分散。在此时期的海洋经济研究主要集中在海洋渔业和海洋运输产业之中，对其他产业较少涉猎。但通过对其具体研究目标的分析可以发现，成果散布于这两

---

① ［苏］A. 柯拉西尔什科夫：《法国的海运业》，载《世界经济文汇》1957年第8期。
② ［挪］小孔尔：《战后挪威海运业的动向》，载《世界经济文汇》1957年第10期。
③ 《水产科技情报》编辑组：《朝鲜民主主义人民共和国海洋渔业》，载《水产科技情报》1973年第6期。
④ 《水产科技情报》编辑组：《苏将通过海洋渔业进一步掠夺世界水产资源》，载《水产科技情报》1973年第2期。

个产业的众多领域之中,但并没有明显的集中研究趋势。如对海洋渔业的论述中就包括了对海洋渔业发展必要性的研究,① 对海洋渔业企业经济体制的研究,② 对渔业生产组织的研究,③ 对渔业技术经济的研究,④ 对渔业金融贷款的研究,⑤ 等等。而在对海洋运输业的研究中也包括了对我国海运产业宏观状况的总结,⑥ 对港口产业布局原则的研究,⑦ 对港口微观经济学的研究,⑧ 等等。

第三,从研究方法看,以对海洋经济活动的现状论述为主要方式,缺乏系统的经济学理论作为支撑,甚至从某种意义上讲,大部分成果都不能说是运用经济学理论进行分析,而只能在实用主义基础上,运用众多相关学科的知识来阐述和解释一个经济问题,不具有经济学理论研究的一般特性。

四、1978~1990年海洋经济初步形成阶段的发展特点

1978年,我国学术界第一次提出了"海洋经济"这个名词,此后围绕它开展了大量的学术活动,海洋经济学科正式产生。此后,在系统海洋经济研究的推动下,国家对于海洋经济的重视程度愈加显著,也越来越为国家高层领导所重视。1981年和1982年,中国海洋国际问题研究会在国家海洋局和中国社会科学院的支持下,组织召开了两次包括"海洋经济"在内的讨论会,并将第一次讨论会有关海洋经济的论文进行了整理,形成了论文集《中国海洋经济研究》,⑨ 这标志海洋经济理论研究的开始。整个20世纪80年代成为我国海洋经济研究大发展的阶段。此时期的海洋经济研究主要进行了三项工作:

(一)完成了海洋经济学体系的初步构建

在一个社会科学学科初步构建之时,必然需要对其研究对象、研究内容、研究目的以及研究方法等进行初步构建,明确该学科与相关学科的特点与差

---

① 王建悌:《领导群众垦殖海洋》,载《中国水产》1958年第9期。
② 马敬通:《十年来国营海洋企业》,载《中国水产》1959年第19期。
③ 广东省湛江专署水产局:《关于海洋捕捞渔业社队经营管理的几个问题的探讨》,载《中国水产》1964年第8期。
④ 夏世福:《渤、黄、东海渔轮技术经济分析》,载《海洋水产研究丛刊》1963年。
⑤ 章道真:《收回渔业贷款的几种做法》,载《中国金融》1959年第15期。
⑥ 张达广:《新中国海运地理十年》,载《陕西师范大学学报》1960年第1期。
⑦ 陈汉欣、林幸青:《我国港口布局原则的初步探讨》,载《中山大学学报》1960年。
⑧ 钟海运:《国外主要港口对CF和CIF价格条件的解释和运用》,载《国际贸易问题》1976年。
⑨ 张海峰:《中国海洋经济研究》,海洋出版社1984年版。

异。因此，在整个20世纪80年代对海洋经济研究体系的探讨贯彻始终，成为当时海洋经济研究的一大主题。

在早期，这种发展趋势主要表现为国外系统海洋经济学研究成果的大规模引入和介绍。实际上，早在20世纪60年代，世界上的主要发达国家（如美国、苏联、日本、法国、加拿大、英国、荷兰和韩国等国是较早对海洋经济进行研究的国家），就相继建立了海洋研究团体和机构，围绕海洋部门经济、海洋区域经济、海洋开发战略和管理等领域展开了多方位的研究，取得了一批重要成果。因此，当我国学术界开始摆脱"文革"所带来的影响，意图建立我国的海洋经济学科时，必然首先需要将目光集中于国外的先进研究成果之上。在此背景下，在20世纪70年代末、80年代初我国的专业经济研究期刊和文章集中出现了一批源引自西方主流经济学刊物的海洋经济学论文（见表1）。这些文章均对当时世界海洋经济学研究的主要范式、方法和框架作了系统论述，为我国海洋经济理论的产生奠定了基础。

表1　1978~1980年我国源引自西方主流经济学刊物的海洋经济学论文

论文题目	作　者	期　刊	年份
海洋经济学、环境、存在问题和经济分析	[美] 莫利斯·威尔金森	国外社会科学文摘	1980
海洋资源美国海洋政策经济学	[美] 詹姆斯·克拉奇菲尔德	国外社会科学文摘	1980
苏联集团海洋政策经济学	[苏] 弗拉迪米尔·卡佐恩斯基	国外社会科学文摘	1980
1985年空间海洋学的状况和展望	[美] D. J. 贝克	海洋开发与管理	1986

在引入西方海洋经济研究体系的同时，1982年中国海洋经济研究会成立。此时期，促使学者也从我国特有的经济环境、学术背景出发，围绕中国海洋经济学的构建以及海洋开发战略、规划和政策等方面开展系统的研究。涌现出了一批具有理论创新意义的文献（见表2），并先后出版了《中国海洋经济研究》（1~3辑）、《中国海洋经济研究大纲》、《海洋经济研究文集》、《中国海洋区域经济研究》和《海岸带管理与开发》等著作。1986年山东社会科学院承担的"中国海洋区域经济研究"是国家"七五"时期社会科学基金项目，也是我国社会科学中第一次系统研究海洋经济的课题。20世纪90年代的这些成果既包括海洋经济理论性问题的探讨，也包括对于其具体分支或领域的基础性研究，从理论高度构建了海洋经济学发展、研究的基础和框架。

表 2　　　　　　　　　1978~1980 年我国海洋经济理论性研究论文

论文题目	作　者	期刊/论文集	年份
海洋、海洋经济与海洋经济学	蒋铁民	中国海洋经济研究第一编	1981
关于开展海洋经济研究的几个问题	张海峰、杨金森	中国海洋经济研究第一编	1981
略论海洋开发和海洋经济理论的研究	何宏权、程福祜	中国海洋经济研究第二编	1982
关于开展我国海洋经济理论研究的设想	杨克平	社会科学	1984
谈一点我对海洋国土经济学研究的认识	于光远	海洋开发与管理	1984
海洋经济学的研究对象、任务和方法	孙凤山	海洋开发与管理	1985
海洋经济学初探	权锡鉴	东岳论丛	1986
建立海洋开发经济学科学体系初探	张爱诚	东岳论丛	1990

但从海洋经济理论的研究进展来看，受当时生产力发展水平和国内经济学理论研究水平所限，此时期的海洋经济研究也不可避免地存在一些问题。第一，对于海洋经济的内涵、海洋经济的归属、海洋经济的主要理论等基本问题研究得尚不够深入，主要集中于对海洋经济的资源属性的定义与描述，仍旧将其视为陆地经济的原料产地，缺乏独立性。第二，从研究方法上讲，主要是通过对实际经济活动中出现的各种问题进行总结和分析，结合马克思主义政治经济学的一般原理进行定性分析，说服力较弱。第三，受研究力量所限，与当时海洋经济快速发展的趋势相比，海洋经济的理论研究也显得较为缓慢，不能很好地满足海洋经济发展的需要。

（二）进行了开展海洋经济发展所必需的一系列基础性工作

在 20 世纪 80 年代初的海洋经济研究起步阶段，为给其创造必要的发展基础，我国各地涉海科研、统计部门作了一系列基础性工作，具体完成了两件工作。

第一，对海洋经济研究的物质基础、海洋物理资源进行了调查。理论源于实践，而经济学家对于实践活动的观察，主要是通过各种统计资料和调查数据来进行的。1949 年新中国成立初始，我国海洋经济管理建立在行业基础之上，各种海洋管理机构由原来大陆的行业管理向海洋延伸形成，未建立统一管理海洋事务的综合管理部门。尽管在 1964 年 7 月国家海洋局正式成立，但国家赋予国家海洋局的宗旨是负责统一管理海洋资源和海洋环境调查资料的收集、整理和海洋公益服务，目的是把分散的、临时性的协作力量转化为一支稳定的海洋工作力量，而中国的海洋经济资料并没有包括其中，这极大地影响了我国海洋经济研究的发展。

1979~1990 年，由国家海洋信息中心主持，国家科委、海洋局和测绘局等单位参加完成的"中国海岸带和海涂资源综合调查"，有五百多个单位近两万人在35万平方公里的海岸带上取得各种观测数据5 800万个，编写各种报告6 000万字，绘制图件数千个，初步弄清了海洋资源的存量和开发状况。此后，国家又于1987~1992年、1983~1989年先后组织二十多个单位进行了全国海岛资源综合调查、世界大洋多金属结核资源调查，获得了大量基础性数据。1988年海洋出版社出版的《1986年中国海洋年鉴》，其中就包括了部分海洋经济统计资料。1989年国务院赋予国家海洋局的职责中明确提出由国家海洋局"负责海洋统计"的工作，国家海洋局组织开展了《海洋统计指标体系》的研究和前期准备工作，自当年起陆续出版了《中国海洋经济统计年鉴》，为海洋经济研究提供了系统的经济信息资源。至此，我国海洋基础数据工作基本完成。

第二，开辟了一系列相关的学术会议、研究机构和学术交流场所。在20世纪80年代之前，我国并没有专门的海洋经济研究组织与刊物，其相关文献均零散地出现于相关领域的期刊之中。如果一门学科长期没有相应的学术交流活动，无疑会使其在学术界与社会上的接受程度受到影响，从而制约其长远的发展。因此，1981年6月，国家海洋局和中国社会科学院经济研究所在北京联合召开了第一次海洋经济研究座谈会，并开始筹备成立"中国海洋经济研究会"。1982年，在山东省社会科学院设立国家级海洋经济研究所；此后出现了许多海洋经济研究机构，有国家海洋局海洋战略研究所、海洋科技情报研究所（后改为海洋信息中心）经济研究室、辽宁师范大学海洋经济地理研究室，等等。1984年，国家海洋局创办了《海洋开发》杂志，指导海洋经济工作实务和理论研究，为广大海洋经济学者提供交流平台。

（三）对当时我国海洋产业发展面临的一些突出问题进行了研究

由于前期缺乏对于海洋经济发展各领域的基础性研究，此时期海洋经济研究成果主要以实用为主，主要是针对实践活动中所存在的实际问题而进行的专题性研究，其内容往往以特点、问题、原因、措施中一部分或几部分作为基本结构，运用定性分析作出结论，缺乏理论基础。受生产力条件所限，20世纪80年代的海洋经济活动仍以资源开发为主体，以海洋渔业和海洋运输产业为主流，海洋盐业、滨海矿砂开采、海洋石油开发为辅。而文献分布也与此相适应，1978~1990年各学术期刊共有海洋经济相关文献480篇，其中研究海洋渔业与海洋运输业发展的占到80%以上，此外海洋区域经济研究与海洋资源经济研究也占一定比例，分别为10.63%和5.21%，这突出体现了20世纪80年代海洋研究以实效性、政策性为主，缺乏基础理论研究的特点。

具体从各研究领域来看，也同样体现出浓郁的实用主义特色。在渔业方面，主要是围绕着当时我国近海渔业资源枯竭的现实问题，有针对性地对我国渔业资源评价、渔业资源保护、海洋水产养殖发展、远洋渔业捕捞发展、海洋渔业多种经营、渔业产业化等一系列问题作了分析与探讨，形成了一系列专著（见表3）和论文。而在海洋运输发展领域，20世纪80年代是我国海运产业开始大规模推行赶超战略的关键时期，因此，大部分分文献集中于研究我国港口布局与开发、远洋航运产业发展战略以及世界各国海运产业发展脉络，以期对我国海运发展提供一定的参考。主要成果有1990年出版的论文集《港口发展与沿海经济》①及各种学术论文135篇。在海洋区域经济方面，则主要是在国家改革开放、大力发展社会主义经济建设的背景下，各沿海省份、城市对如何利用自身在海洋方面的地理位置与资源优势，制定合理的经济发展战略的研究，主要成果有天津市哲学社会科学学会联合《沿海城市经济研究》编辑组编撰的《沿海城市经济研究》论文集，以及蒋铁民主编的《中国海洋区域经济研究》。在海洋资源开发的研究中，主要研究20世纪80年代已初步具有开发能力的资源，如海洋石油、海洋矿产、海洋能源、滨海矿砂等的经济效果和前景问题，内容较为分散，缺乏综合性、产业化的研究文献。

表3　　　　　　　　　　1978~1990年海洋渔业研究成果

书　名	作　者	出版社	年份
渔业技术经济分析	夏世福	农业出版社	1980
海洋水产资源调查手册	夏世福、刘效舜	科学技术出版社	1981
南海北部大陆坡渔业资源综合考察报告	南海水产研究所	南海水产研究所编印	1981
东海大陆架外缘和大陆坡深海渔场综合调查研究报告	东海水产研究所	东海水产研究所编印	1984
远洋渔业	沈汉祥等	海洋出版社	1987
渔业经济生态学概论	夏世福	海洋出版社	1989
黄、渤海区渔业资源调查和区划	刘效舜、吴敬南等	海洋出版社	1990
山东近海渔业资源开发与保护	唐启升等	农业出版社	1990
中国渔业资源调查和区划	夏世福等	农业出版社	系列专著

总而言之，20世纪80年代是我国海洋经济研究事业从无到有、从小到大的重要阶段。其在整个海洋经济研究发展进程中，起到了无法替代的作用，这个时期初步建立海洋经济的理论体系和研究方法对于提高人们的海洋意识，促进海洋经济的快速、健康发展，鼓励学者对海洋经济理论的深入研究等方面均具有重要意义。

① 杨玉生：《港口发展与沿海经济》，大连海运出版社1990年版。

五、1991~2000年海洋经济研究迅猛发展时期的特点

1991~2000年，海洋经济实践的快速发展和各种西方经济学方法论的引入为我国海洋经济研究提供了实践和思想基础，此时期海洋经济研究开始呈现出大范围、多角度、多元化的研究趋势。此时期我国各学术期刊共刊发海洋经济相关论文980篇，各种著作、文集、报告等资料一百多部，都较上一时期有了大幅度的增长。在国家、政府的高度重视之下，海洋经济研究呈现出迅猛增长的势头，整个海洋经济研究也进入了一个新的阶段，主要呈现出以下特点。

（一）研究的系统性、理论性有了极大的增强

1991~2000年是我国海洋经济大发展的10年，在这期间海洋经济在我国国民经济中的地位不断提高，受此影响海洋经济研究在整个经济研究体系中的意义也不断扩大。海洋经济研究已不仅仅是作为陆地经济的资源产地，如煤炭、石油等资源经济一般的研究领域，许多学者已逐渐将海洋经济扩展到其生产、流通、分配、消费的一系列产业活动中去，将其视为国民经济中的重要部门经济学科或区域经济学科，对其学科的系统性进行了研究。海洋经济学科体系逐步建立，对其体系下的海洋资源经济学、海洋产业经济学、海洋区域经济学、海洋环境经济学的研究，已开始逐步摆脱了单纯的问题—原因—对策的实用主义文章范式，开始运用西方经济学中的产业组织理论、产业增长理论、产权理论、区域位理论、资源禀赋理论、公共品理论等先进的经济学方法论，对其基本原理进行讨论。出现了一批如表4所示的具有极强理论性著作。

表4　　　　　　　　1991~2000年海洋经济理论性研究成果

书　名	作者	出版社	年份
海洋经济学	孙斌、徐质斌	青岛出版社	2000
国际航运经济新论	徐建华	人民交通出版社	1997
现代港口经济学	邹俊善	人民交通出版社	1997
海运经济地理	王晶、唐丽敏	大连海事大学出版社	1999
海洋产业优化模式	郑培迎	海洋出版社	1997
沿海经济学	潘义勇	人民出版社	1994
海洋资源与可持续发展	鹿守本	中国科学技术出版社	1999

具体来看，在这10年中，我国各学术期刊共刊发海洋经济相关论文980篇，其中对于海洋经济理论性研究的文献为257篇，占总数的26.22%，比上一时期的14.5%有大幅提高。从研究领域上看，理论性较强的海洋经济理论研究、海洋区域经济研究、海洋产业理论研究、海洋环境经济研究分别占总比例的8%、10%、7%和3%，较上一时期分别增长了412%、100%、1360%和733%，表现出对这些理论强劲的增长趋势。而与之相对，在我国海洋经济研究中长期占据主要地位的海洋渔业经济研究和海洋运输经济研究则增长不大，仅增长了10.35%和26.63%，在文献总数中的比例也下降到了50%以下，为46.12%。（图2）造成这种趋势的主要原因是海洋经济性理论的研究对这两种传统的产业性经济研究领域有极强的替代性。首先，从严格意义上说，这两种经济研究类型可以算做海洋产业经济中的具体产业经济研究，但由于其在海洋经济发展历程中的特殊地位，在早期不得不将其独立出来。其次，在进行这两种产业的研究时，不可避免地将会运用到海洋经济理论研究所提供的各种方法论进行指导，如海洋产业经济理论可以对海洋渔业产业化和海运产业化的发展趋势构成指导，海洋区域经济理论可以对港口产业的布局提供建议，等等。此外，还值得注意的是对于海洋旅游经济的研究在此时期开始出现，这是与当时的海洋经济开发实践相符合的。

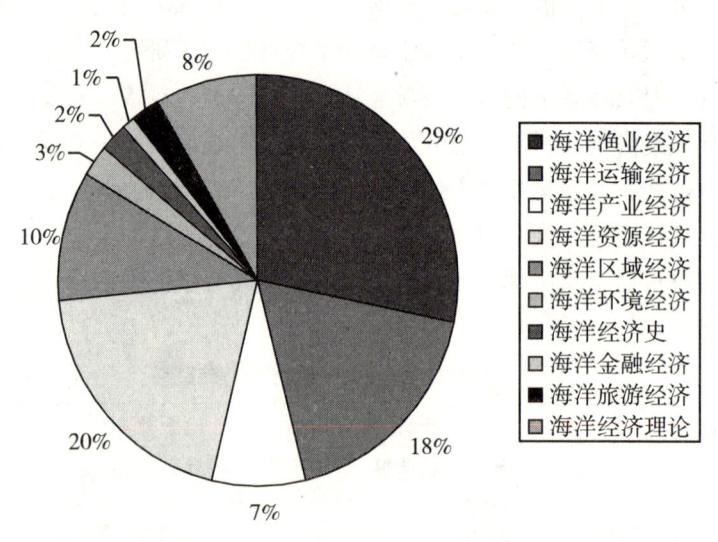

图2　1991~2000年海洋经济研究成果分布示意图

（二）我国海洋经济发展战略研究的高潮

1991年1月，在北京召开了全国首次海洋工作会议，确定了20世纪90年代中国海洋工作的基本指导思想：以开发海洋资源、发展海洋经济为中心，围绕"效益、资源、环境、减灾"主题开展工作。1996年《中国海洋21世纪议程》正式发布，这是中国实施海洋可持续开发利用的政策指南。基本战略原则是以发展海洋经济中心，适度快速开发，海陆一体化开发，科教兴海和协调发展。此后，随着1998年"国际海洋年"的到来，我国开始掀起海洋经济发展研究的高潮，各种合作协议、学术研讨会、知识培训、考察访问活动层出不穷，同时我国学者还广泛参与海洋领域的国际合作与交流活动。在此基础上，许多学者撰文分析了我国发展海洋经济的重要性，提出了"海上中国"的说法。此后他们围绕政治、主权、国家安全、经济发展、资源开发等领域，开展了一系列战略性研究（见表5）。

表5　　　　　1991~2000年中国海洋经济发展战略研究成果

书　名	作　者	出版社	年份
建设海上中国纵横谈	王诗成	山东友谊出版社	1995
中国海洋开发与管理	刘洪滨	香港天马图书公司	1996
中国海洋21世纪议程	国家海洋局	海洋出版社	1996
中国海洋政策	国家海洋局海洋发展战略研究所	海洋出版社	1998
海洋资源开发与管理	陈学雷	科学出版社	2000
建设海洋经济强国方略	王诗成	泰山出版社	2000
海洋强国论	鹿守本	海洋出版社	1999

在国家海洋经济发展战略的研究热潮之下，我国沿海各省市也出现"蓝色国土"开发热，各级政府职能部门对于海洋经济的热情不断高涨。在此推动之下，众多新的海洋经济研究机构和海洋相关院校得以成立。1997年湛江海洋大学开始设立了海洋经济学专业，开始培养海洋经济方面的专门人才，全国海洋经济研究队伍得到了壮大。在沿海各省份的大力倡导之下，产生了一系列针对各省海洋经济发展的研究成果。"海上山东"、"海上辽宁"、"海上浙江"、"海上广东"等说法屡见不鲜，甚至还出现了"海上苏东"、"海上锦州"、"海上连云港"的理论，此外各种"海洋强（大）省"、"海洋活县（市）"等论述更是汗牛充栋。这充分体现了，在社会主义市场经济条件下，出于地方经济增长的考虑，当时的地方各级政府对于海洋经济的极高热情。因此，这些研究主要是从海洋经济发展和地区的资源禀赋特点出发，对地方各省、市、县的海洋相关产业发展进行了战略性研究，形成了一系列"战略性构想"、"发展规划"、"综合开发"、"发展方向"、"经济跳跃发展"之类的研究，取得了一系列成果（见表6）。

表6　　　　　　1991~2000年区域海洋经济发展战略研究成果

书　名	所属省份	作　者	出版社	年份
建设海上山东	山东	鞠茂勤	海洋出版社	1992
海洋经略	广西	郑白燕	广西科技出版社	1992
山东海洋经济	山东	孙义福	山东人民出版社	1994
希望在海洋	江苏	江苏省海洋管理局	江苏人民出版社	1994
建设海洋大省	福建	福建省计划委员会	鹭江出版社	1996
建设海上山东战略措施研究	山东	高洪涛、徐质斌	海洋出版社	1996
加速海上山东建设研究	山东	黄学军	海洋出版社	1997
长山群岛经济社会系统分析——辽宁省长海县综合发展战略研究	辽宁	张耀光、张云瑞	辽宁师范大学出版社	1997
山东海洋产业结构和布局优化研究	山东	徐质斌、孙吉亭	海洋出版社	1998
海上山东建设概论	山东	郑贵斌、徐质斌	海洋出版社	1998
河北省海洋经济发展研究	河北	尹紫东	海洋出版社	1998
广东海洋经济	广东	王荣武、梁松	广东人民出版社	1998
辽宁海岛资源开发与海洋产业布局	辽宁	张耀光、胡宜鸣	辽宁师范大学出版社	1998
浙江建设海洋经济大省战略研究	浙江	苏纪兰、蒋铁民	海洋出版社	1999
21世纪的粤湛经济：海洋知识经济	广东	叶远谋	当代中国出版社	2000
海南经济经济特区定位研究	海南	李克	海南出版社	2000

六、21世纪初海洋经济研究成熟阶段的特点

21世纪的海洋经济研究成果也十分丰富，截至2007年，共刊发海洋经济相关各种论文1 751篇，比上一时期增加了78.67%，体现出了较前一时期更为迅猛的增长势头。这主要是由于在21世纪初，我国大规模的海洋经济发展已经历了二十多年发展的历程，随着实践活动的深入，海洋经济的研究范畴进一步扩大，如陈可文（2003）就认为广义上的海洋经济可以包括"与海洋经济难以分割的海岛上的陆域产业海岸带的陆域产业及河海体系中的内河经济等，包括海岛经济和沿海经济"。这在一定程度上将海洋经济独立于陆域经济，形成独立的经济系统。因此，学者的研究重点也逐步由对海洋开发利用中的局部问题研究上升到全局问题研究，由对海洋经济发展的现实对策研究提升到运用经济理论对海洋经济理论基本框架、体系等原理问题的探讨。从一定意义上可以说，研究方向、研究内容、研究形式的转变，标志着海洋经济理论研究的

质的飞跃。至此，海洋经济研究逐渐形成了自己独立的研究范式和框架，理论与实证相结合的研究方法逐步确立。这种成熟的研究氛围，必定会促使更多的学者加入海洋经济研究的行列。与之相适应，此时期的海洋经济研究也呈现出了新的时代特点。

（一）以可持续发展为核心的新研究领域不断发展、壮大

从 21 世纪的海洋经济研究范畴看，绝大部分传统研究领域增长较为平缓，如海洋渔业经济研究、海洋产业经济研究、海洋区域经济研究、海洋经济史研究、海洋经济理论研究分别增长了110%、102%、93.1%、80%和58.5%，基本与此时期的整体增长趋势相符。此外也有一些研究领域陷入停滞，如海洋运输经济研究、海洋资源经济研究，仅分别增长了 7.42% 和 7.77%，这是由这两种学说所对应研究的领域在海洋经济实践活动中的地位下降造成的（见图3）。

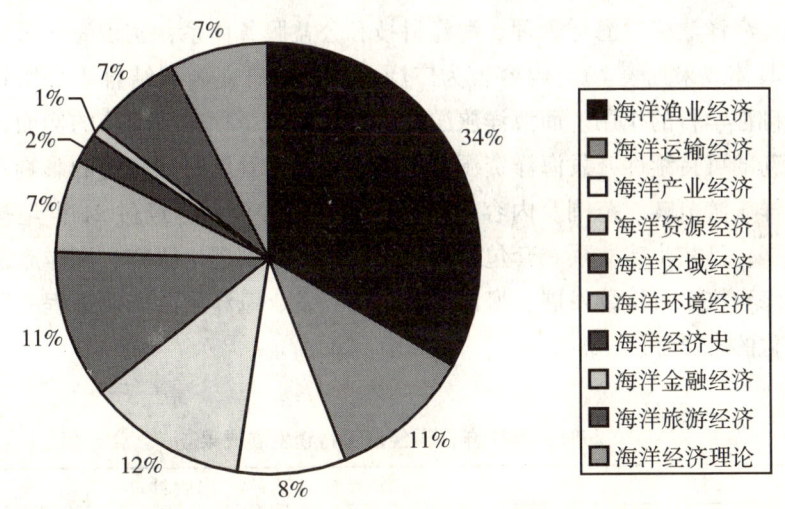

图3　21世纪以来海洋经济研究成果分布示意图

在这种背景下，特别值得注意的是，在 20 世纪 90 年代新兴的海洋经济研究领域，即海洋旅游经济研究和海洋环境经济研究，在此时期有了长足的发展，分别较前一时期增长了 388% 和 459%，在总量中已分别占到 7% 的比例。海洋环境经济和海洋旅游业的研究，从实质上讲都是从可持续开发的角度对海洋资源的运用研究，这是 21 世纪经济发展对于海洋资源开发所提出的新要求。海洋经济可持续发展是一种新的发展观，是指以人类社会与自然和谐发展为目标，以经济社会与环境协调为途径，逐步实现人口、环境、资源与发展的协

调。对于海洋开发同样要做到生态系统、经济效益和社会公平三者协调发展。做到人、海、地域经济系统在发展上的和谐，使海洋资源得以持续利用，海洋经济持续增长，海陆经济共同发展（高强，2006）。

实际上我国提出海洋可持续发展理论研究的历史很长，从20世纪80年代中期就开始追踪国际相关动向。并先后在1994年和1996年颁布了《中国21世纪议程》和《中国海洋21世纪议程》，提出中国海洋事业要走可持续发展的道路。但直到2001年，受经济发展条件所限，学者对可持续发展的意识并不强烈。虽然许多海洋经济研究相关论文均提到海洋可持续发展问题，但专门针对该问题的理论研究成果并不多。2001年国家海洋局印发了《海洋工作"十五"计划纲要》，提出了以满足社会经济发展对海洋资源不断增长的需要为基本出发点，推动海洋经济可持续发展的主题，极大地推动了海洋可持续发展研究的进程。从具体研究成果来看，海洋环境经济研究主要侧重于研究21世纪以来的海洋生态环境保护问题，研究方向主要有：合理开发海洋资源，维护我国海洋权益，强化海域资源管理，保证海洋环境安全，建立和完善海洋开发规划、海洋法律、海洋管理、海洋科技和公益服务体系，实现海洋可持续发展等（具体成果见表7），内容较为广泛，与相关研究领域结合十分密切，起到了基础性学科的作用。而海洋旅游经济则是从另一种产业经济的侧面，提出了一条新的可持续性开发海洋资源的道路。虽然受发展时间较短的影响，目前对于海洋旅游范畴、范围及内容仍然模糊并存在异议，但经过21世纪初六七年的发展，已初步建立了一套包括海洋旅游经济基础理论研究、海洋旅游资源开发、海洋旅游可持续发展、海洋区域旅游开发、海洋旅游产业进程一系列领域系统化的海洋经济学科。

表7　　　　　　　　2001～2007年海洋经济可持续发展成果

书　名	作者	出版社	年份
山东海洋资源与环境	李荣升	海洋出版社	2002
海岸带可持续发展与综合管理	恽才兴	海洋出版社	2002
厦门市海洋经济发展战略和海洋环境保护研讨会论文集	厦门市海洋与渔业局	海洋出版社	2006
山东省滨海旅游及旅游业	刘洪滨	海洋出版社	2004
海洋旅游学	董玉明	海洋出版社	2007

（二）海洋经济学研究的交叉化趋势

进入21世纪的中国海洋经济研究，学科体系已经确立，在理论研究上已达到较为完善的阶段,对于各个领域的基础性理论和热点问题研究,在现有方法论条件下,已达到一个较为完善的程度,但尚不具备高速发展的空间。因此,此时期许多学者开始转向对学科交叉性的研究。这种交叉既有海洋经济研究体系之内的内涵性交叉,也包括海洋经济理论与其他学科门类的外延性交叉。

内涵性交叉研究倾向是指学术界开始将目光转向原来并不为学界所重视的一些较为细致、具有多种海洋经济研究领域交叉属性的问题之中(表8列出了这种交叉趋势的典型成果)。这是由海洋经济学理论的供给和需求两方面作用引发的。从需求角度讲,21世纪的海洋开发实践活动,无论是从规模上还是深度上,都达到了前所未有的高度。因此,在实践中所遇到了各种问题也愈加复杂,涉及的因素众多,很难用单一化的理论进行研究。这也对海洋经济研究的交叉化提出了要求。而从供给方面讲,这种学科交叉的趋势是海洋经济学理论研究进入系统化发展阶段的必要条件。这种交叉性研究成果的形成,是来源于海洋经济学领域各基础性研究课题的集合,又是各研究领域专业分工的必然结果。这种研究领域的分工帮助学者对特定领域能够更为深入地进行研究,提高其专业素养;但另一方面,学者在面临其他相关研究领域的问题时,也会习惯性地运用自身较为熟悉的方法论来解决,这无疑将促使不同研究领域之间的知识点相互作用,形成新的研究点。

表8　　　　　　2001~2007年海洋经济学研究交叉化趋势典型成果

文章题目	学科属性	年份
黄渤海区海洋渔业可持续发展研究	环境经济/渔业经济	2006
中国海洋渔业资源可持续利用和有效管理研究	环境经济/渔业经济	2006
论环境资源制约下我国海洋产业结构的优化策略	产业经济/环境经济	2006
海洋经济强省建设下山东国际航运发展战略研究	区域经济/产业经济/运输经济	2006
区域海洋产业合理布局的问题及对策	产业经济/区域经济	2004
山东省海洋旅游业可持续发展系统分析与评价	环境经济/区域经济/旅游经济	2005
浙江海洋旅游产业区位分析及空间重构	旅游经济/区域经济	2006
保护海岛资源科学开发和利用海岛	资源经济/区域经济/环境经济	2004
国民经济核算体系与海洋生态资源核算	宏观经济/资源经济/环境经济	2004
海岛地区产业演替及资源基础分析	资源经济/区域经济/产业经济	2005
休闲海钓及其发展趋势	旅游经济/渔业经济	2006
建设广西渔业区域经济的战略思考	区域经济/渔业经济	2002

海洋经济学外延上的交叉性主要体现在与其他相关学科的交叉性研究之上,此时期海洋经济外延上的扩展主要体现于与社会人文学科的融合之上。海

洋经济学作为一门资源性特征很强的经济学科自产生之时起，就与相关的海洋自然学科有着千丝万缕的联系，20世纪80年代较为盛行的海洋技术经济研究和对资源开发的经济效果研究便是典型的例子。到了21世纪，随着海洋事业整体水平的发展，与之伴生的海洋人文科学弱势已经日益明显地凸显出来。海洋文化在海洋经济活动中产生、形成和发展，而海洋经济发展则需要海洋文化发挥基础作用，为其提供智力和活力支持。因此，从20世纪90年代末期开始，许多学者运用其他社会学科方法论和经济学理论结合，对海洋经济活动进行研究。他们往往在经济分析中引入历史学、社会学、心理学、宗教学的理论，从海洋社会经济史和海洋人文社会的视野对海洋经济活动进行考察，产生了一批较有意义的成果，比较典型的如江西高校出版社出版的《海洋与中国丛书》（共两辑），海洋出版社出版的《蓝色的畅想》（李向民，2007），中山大学出版社出版的《海南经济史》（陈光良，2004），齐鲁书社出版的《山东沿海开发史》（王赛时，2005），等等。

　　海洋经济交叉研究的发展，是海洋经济发展成熟的体现。它极大地扩展了海洋经济学科知识的覆盖面，形成一条由早期的点——单个热点海洋经济问题研究（20世纪80年代左右），到线——如海洋区域经济学、海洋产业经济学、海洋运输经济学等专业性、理论性研究领域的兴起（20世纪90年代左右），再到面——各理论性研究成果形成内涵性的交叉（21世纪）——最后与其他面（其他学科）构成海洋经济研究的空间体系（21世纪以后），最后组合成涵盖海洋经济活动各个方面，包括理论研究和应用研究系统的完整经济学发展脉络，完成了学科体系的最终构建。而这种完成并不是意味着海洋经济学可以高枕无忧了，恰恰相反，完善的海洋经济学体系正是给予海洋经济学者以更为强大的方法论武器，使之可以面对未来更为复杂的海洋经济实践活动，不断丰富海洋经济学的内涵，扩展其外延，促进海洋经济活动与海洋科学理论的不断进步。

参考文献

[1] D.J.贝克.1985年空间海洋学的状况和展望[J].海洋开发与管理，1986（3）.

[2] 威尔金森.海洋经济学：环境，存在问题和经济分析[J].国外社会科学文摘，1980（9）.

[3] 詹姆斯·克拉奇菲尔德.海洋资源——美国海洋政策经济学[J].国外社会科学文摘，1980（9）.

[4] 小孔尔.战后挪威海运业的动向[J].世界经济文汇，1957（10）.

[5] A.柯拉西尔什科夫.法国的海运业[J].世界经济文汇，1957（8）.

[6] 弗拉迪米尔·卡佐恩斯基. 苏联集团海洋政策经济学 [J]. 国外社会科学文摘, 1980 (9).

[7] 曹忠祥, 等. 区域海洋经济发展的结构性演进特征分析 [J]. 人文地理, 2005 (12).

[8] 陈东有. 走向海洋贸易带: 近代世界市场互动中的中国东南商人行为 [M]. 南昌: 江西高校出版社, 1998.

[9] 陈光良. 海南经济史 [M]. 广州: 中山大学出版社, 2004.

[10] 陈汉欣, 林幸青. 我国港口布局原则的初步探讨 [J]. 中山大学学报, 1960.

[11] 陈可文. 中国海洋经济学 [M]. 北京: 海洋出版社, 2003.

[12] 陈万灵. 关于海洋经济的理论界定 [J]. 海洋开发与管理, 1998 (3).

[13] 陈万灵. 海洋经济学理论体系的探讨 [J]. 海洋开发与管理, 2001 (5).

[14] 陈学雷. 海洋资源开发与管理 [M]. 北京: 科学出版社, 2000.

[15] 东海水产研究所. 东海大陆架外缘和大陆坡深海渔场综合调查研究报告 [R]. 东海水产研究所内部资料, 1984.

[16] 董伟. 美国海洋经济相关理论和方法 [J]. 海洋经济, 2005 (4).

[17] 董玉明. 海洋旅游学 [M]. 北京: 海洋出版社, 2007.

[18] 福建省计划委员会. 建设海洋大省 [M]. 厦门: 鹭江出版社, 1996.

[19] 高洪涛, 徐质斌. 建设海上山东战略措施研究 [M]. 北京: 海洋出版社, 1996.

[20] 广东省湛江专署水产局. 关于海洋捕捞渔业社队经营管理的几个问题的探讨 [J]. 中国水产, 1964 (8).

[21] 国家海洋局. 海洋工作"十五"计划纲要 [R]. 内部资料, 2001.

[22] 国家海洋局. 中国海洋21世纪议程 [M]. 北京: 海洋出版社, 1996.

[23] 国家海洋局海洋发展战略研究所. 中国海洋政策 [M]. 北京: 海洋出版社, 1998.

[24] 韩立民, 王爱香. 保护海岛资源科学开发和利用海岛 [J]. 海洋开发与管理, 2004 (6).

[25] 何翔舟. 我国海洋经济研究的几个问题 [J]. 海洋科学, 2002 (1).

[26] 黄顺力. 海洋迷思: 中国海洋观的传统与变迁 [M]. 南昌: 江西高校出版社, 1999.

[27] 黄学军. 加速海上山东建设研究 [M]. 北京: 海洋出版社, 1997.

[28] 江苏省海洋管理局. 希望在海洋 [M]. 南京: 江苏人民出版社, 1994.

[29] 蒋铁民. 中国海洋区域经济研究 [M]. 北京: 海洋出版社, 1990.

[30] 鞠茂勤．建设海上山东［M］．北京：海洋出版社，1992．

[31] 蓝达居．喧闹的海市——闽东南港市兴衰［M］．南昌：江西高校出版社，1999．

[32] 李克．海南经济经济特区定位研究［M］．海口：海南出版社，2000．

[33] 李荣升．山东海洋资源与环境［M］．北京：海洋出版社，2002．

[34] 李向民．蓝色的畅想［M］．北京：海洋出版社，2007．

[35] 联合国经济及社会理事会海洋经济技术处．海岸带管理与开发［M］．北京：海洋出版社，1988．

[36] 刘洪滨．山东省滨海旅游及旅游业［M］．北京：海洋出版社，2004．

[37] 刘洪滨．中国海洋开发与管理［M］．香港：香港天马图书公司，1996．

[38] 刘效舜，吴敬南．黄、渤海区渔业资源开发与保护［M］．北京：海洋出版社，1990．

[39] 刘正刚．东渡西进：清代闽粤移民台湾与四川的比较［M］．南昌：江西高校出版社，2004．

[40] 楼东，谷树忠，朱兵见，刘盛和．海岛地区产业演替及资源基础分析［J］．经济地理，2005（4）．

[41] 卢秀容．中国海洋渔业资源可持续利用和有效管理研究［D］．武汉：华中农业大学，2005．

[42] 卢兆发．建设广西渔业区域经济的战略思考［J］．海洋渔业，2002（1）．

[43] 鹿珑．黄渤海区海洋渔业可持续发展研究［D］．天津：天津大学，2006．

[44] 鹿守本．海洋强国论［M］．北京：海洋出版社，1999．

[45] 鹿守本．海洋资源与可持续发展［M］．北京：中国科学技术出版社，1999．

[46] 马敬通．十年来国营海洋企业［J］．中国水产，1959（19）．

[47] 马丽卿．浙江海洋旅游产业区位分析及空间重构［J］．浙江学刊，2006（4）．

[48] 南海水产研究所．南海北部大陆坡渔业资源综合考察报告［R］．内部资料，1981．

[49] 欧阳宗书．海上人家：海洋渔业经济与渔民社会［M］．南昌：江西高校出版社，1998．

[50] 潘义勇．沿海经济学［M］．北京：人民出版社，1994．

[51] 全锡鉴．海洋经济学初探［J］．东岳论丛，1986（6）．

[52] 沈汉祥．远洋渔业［M］．北京：海洋出版社，1987．

[53] 石洪华, 郑伟, 丁德文, 高会旺, 刘洋. 关于海洋经济若干问题的探讨 [J]. 海洋开发与管理, 2007 (1).

[54] 水产科技情报编辑组. 朝鲜民主主义人民共和国海洋渔业 [J]. 水产科技情报, 1973 (6).

[55] 水产科技情报编辑组. 苏将通过海洋渔业进一步掠夺世界水产资源 [J]. 水产科技情报, 1973 (2).

[56] 苏纪兰, 蒋铁民. 浙江建设海洋经济大省战略研究 [M]. 北京: 海洋出版社, 1999.

[57] 孙斌, 徐质斌. 海洋经济学 [M.]. 青岛: 青岛出版社, 2000.

[58] 孙凤山. 海洋经济学的研究对象、方法和任务 [J]. 海洋开发与管理, 1985 (5).

[59] 孙义福. 山东海洋经济 [M]. 济南: 山东人民出版社, 1994.

[60] 唐启升. 山东金海渔业资源开发与保护 [M]. 北京: 农业出版社, 1990.

[61] 天津市哲学社会科学学会联合会,《沿海城市经济研究》编辑组. 沿海城市经济研究 [R]. 内部资料, 1982.

[62] 王建悌. 领导群众垦殖海洋 [J]. 中国水产, 1958 (9).

[63] 王晶, 唐丽敏. 海洋经济地理 [M]. 大连: 大连海事大学出版社, 1999.

[64] 王淼, 刘晓洁, 李洪田. 国民经济核算体系与海洋生态资源核算 [J]. 乡镇经济, 2004 (11).

[65] 王琪, 何广顺, 高忠文. 构建海洋经济学理论体系的基本设想 [J]. 海洋信息, 2005 (3).

[66] 王荣国. 海洋神灵: 中国海神信仰与社会经济 [M]. 南昌: 江西高校出版社, 2003.

[67] 王荣武, 梁松. 广东海洋经济 [M]. 广州: 广东人民出版社, 1998.

[68] 王赛时. 山东沿海开发史 [M]. 济南: 齐鲁书社, 2005.

[69] 王诗成. 建设海上中国纵横谈 [M]. 济南: 山东友谊出版社, 1995.

[70] 王诗成. 建设海洋经济强国方略 [M]. 济南: 泰山出版社, 2000.

[71] 王震. 山东省海洋旅游业可持续发展系统分析与评价 [D]. 青岛: 中国海洋大学, 2007.

[72] 夏世福, 刘效舜. 海洋水产资源调查手册 [M]. 北京: 科学技术出版社. 1981.

[73] 夏世福. 渤、黄、东海渔轮技术经济分析 [J]. 海洋水产研究丛刊,

1963.

[74] 夏世福. 渔业技术经济分析 [M]. 北京：农业出版社, 1980.

[75] 夏世福. 渔业经济生态学概论 [M]. 北京：海洋出版社, 1989.

[76] 夏世福. 中国渔业资源调查与区划 [M]. 北京：农业出版社, 系列专著.

[77] 厦门市海洋与渔业局. 厦门市海洋经济发展战略和海洋环境保护研讨会论文集 [G]. 北京：海洋出版社, 2006.

[78] 徐建华. 国际航运经济新论 [M]. 北京：人民交通出版社, 1997.

[79] 徐杏. 海洋经济理论的发展与我国的对策 [J]. 海洋战略, 2002 (2).

[80] 徐质斌, 牛福增. 海洋经济学教程 [M]. 北京：经济科学出版社, 2003.

[81] 徐质斌, 孙吉亭. 山东海洋产业结构和布局优化研究 [M]. 北京：海洋出版社, 1998.

[82] 徐质斌. 海洋经济研究文集 [G]. 山东社会科学院海洋经济研究所内部资料, 1989.

[83] 徐质斌. 海洋经济与海洋经济科学 [J]. 海洋科学, 1995 (5).

[84] 许小江. 休闲海钓及其发展趋势 [J]. 浙江海洋学院学报（自然科学版）, 2006 (4).

[85] 杨国桢. 东溟水土：东南中国的海洋环境与经济开发 [M]. 南昌：江西高校出版社, 2003.

[86] 杨国桢. 闽在海中：追寻福建海洋发展史 [M]. 南昌：江西高校出版社, 1998.

[87] 杨海军. 国外发展海洋经济的理论与实践 [J]. 浙江经济, 2005 (12).

[88] 杨克平. 关于开展我国海洋经济理论研究的设想 [J]. 社会科学, 1984 (9).

[89] 杨强. 北洋之利——古代渤黄海区域的海洋经济 [M]. 南昌：江西高校出版社, 2003.

[90] 杨玉生. 港口发展与沿海经济 [M]. 大连：大连海运出版社, 1990.

[91] 叶向东. 现代海洋经济理论 [M]. 北京：冶金工业出版社, 2006.

[92] 叶远谋. 21世纪的粤湛经济：海洋知识经济 [M]. 北京：当代中国出版社, 2000.

[93] 尹紫东. 河北省海洋经济发展研究 [M]. 北京：海洋出版社, 1998.

[94] 于光远. 谈一点我对海洋国土经济学研究的认识 [J]. 海洋开发与管理, 1984 (1).

[95] 于永海, 苗丰民, 张永华, 等. 区域海洋产业合理布局的问题及对策

[J]．国土与自然资源研究，2004（1）．

[96] 于运全．海洋天灾——中国历史时期的海洋灾害与沿海社会经济 [M]．南昌：江西高校出版社，2005．

[97] 恽才兴．海岸带可持续发展与综合管理 [M]．北京：海洋出版社，2002．

[98] 曾少聪．东洋航路移民：明清海洋移民台湾与菲律宾的比较研究 [M]．南昌：江西高校出版社，1998．

[99] 张爱诚．建立海洋开发经济学科学体系初探 [J]．东岳论丛，1990（5）．

[100] 张彩霞．海上山东：山东沿海地区的早期现代化历程 [M]．南昌：江西高校出版社，2004．

[101] 张达广．新中国海运地理十年 [J]．西安：陕西师范大学学报，1960（1）．

[102] 张德贤．海洋经济可持续发展理论研究 [M]．青岛：青岛海洋大学出版社，2000．

[103] 张海峰，张立新．《海洋经济学》评介 [J]．海洋开发与管理，2000（2）．

[104] 张海峰．中国海洋经济研究，第一辑 [M]．北京：海洋出版社，1984．

[105] 张海峰．中国海洋经济研究，第二辑 [M]．北京：海洋出版社，1984．

[106] 张海峰．中国海洋经济研究，第三辑 [M]．北京：海洋出版社，1986．

[107] 张海峰．中国海洋经济研究大纲 [M]．北京：海洋出版社，1986．

[108] 张卫国．海洋经济强省建设下山东国际航运发展战略研究 [D]．青岛：中国海洋大学，2007．

[109] 张晓宁．天子南库——清前期广州制度下的中西贸易 [M]．南昌：江西高校出版社，1999．

[110] 张耀光，胡宜鸣．辽宁海岛资源开发与海洋产业布局 [M]．沈阳：辽宁师范大学出版社，1998．

[111] 张耀光，张云瑞．长山群岛经济社会系统分析——辽宁省长海县综合发展战略研究 [M]．沈阳：辽宁师范大学出版社，1997．

[112] 章道真．收回渔业贷款的几种做法 [J]．中国金融，1959（15）．

[113] 郑白燕．海洋经略 [M]．南宁：广西科技出版社，1992．

[114] 郑贵斌、徐质斌．海上山东建设概论 [M]．北京：海洋出版社，1998．

[115] 郑贵斌．海洋经济位理论与海洋经济创新发展 [J]．海洋开发与管理，2006（5）．

[116] 郑培迎．海洋产业优化模式 [M]．北京：海洋出版社，1997．

[117] 中国21世纪议程管理中心. 中国21世纪议程[M]. 北京: 中国环境科学出版社, 1994.

[118] 钟海运. 国外主要港口对CF和CIF价格条件的解释和运用[J]. 国际贸易问题, 1976.

[119] 周罡. 论环境资源制约下我国海洋产业结构的优化策略[D]. 青岛: 中国海洋大学, 2006.

[120] 周江. 海洋经济发展探析[J]. 天府新论, 2002 (6).

[121] 邹俊善. 现代港口经济学[M]. 北京: 人民交通出版社, 1997.

Research on the evolution of the theories of marine economics in China

Jiang Xuzhao, Huang Cong

[Abstract] Based on the theory essence of marine economics, firstly, this thesis discusses the overall law of evolution on the theories of marine economics since founding of new china and different characteristics of every period; secondly, analyses the development of its definition, methodology and the structure of theoretical system; then, summarizes the general law of its evolution, i. e. systematic evolution process of point-line-plane-space. Finally, there will be an expectation of its developing trend in future.

[Key Words] Marine economics   History of economic theories   Evolution

JEL Classification: B29, Q56, N55

国际海洋保护区研究进展：
一个经济学视角

刘 康[*]

【摘要】 海洋保护区作为一种预防性的海洋综合管理工具，正在逐渐成为沿海各国应对海洋环境污染、生物多样性丧失、资源衰退及生境丧失等海洋生态系统压力的重要手段。海洋保护区发展具有生态及社会环境双重目的，其社会经济价值是决定海洋保护区成败的关键因素，也是海洋保护区研究的重点内容之一。无论是实证研究结果，还是理论模型都证明了海洋保护区的生态价值，但其渔业及游憩业等经济价值却依然存在争议。通过对现有的海洋保护区相关研究文献进行综合分析，发现海洋保护区具有长期的渔业及游憩价值，但也存在短期的渔业机会成本。由于实证分析的困难，现有的多数海洋保护区相关经济学研究都建立在种群生物学模型基础上，包括集合种群模型、空间经济学模型、确定性模型及随机模型等多种生物经济学理论模型，其中多数理论模型研究结果证明了海洋保护区的经济价值，但具有较高的变化性及不确定性。

【关键词】 海洋保护区　经济价值　生物经济学模型　成本效益分析

随着世界各地海洋开发活动的深入发展，人类已经从根本上改变了近海海洋生态系统，海洋生态系统的完整性和稳定性正在受到人类社会过度利用的威胁，这与我们传统的海洋环境认知形成鲜明对比（Agardy，2001）。据美国

---

[*] 刘康，山东海洋经济研究中心副研究员，青岛：266071，Email：kanglk@qingdaonews.com。

《科学》杂志报道，人类对海洋的影响远远超出人们的预想，全球41%的海域已经受到多种人类活动的强烈影响，海洋正面临着前所未有的生态压力（Halpern et al, 2008）。海洋环境污染、生物多样性丧失、资源衰退以及生态环境（生境）丧失已成为沿海地区所普遍面临的问题。自20世纪中叶以来，以格劳修斯为代表的"海洋自由论"面临越来越大的困境，人们开始从"海洋自由"走向"海洋保护"（Russ & Zeller, 2003），海洋保护区成为人类保护海洋环境，持续利用海洋资源的一种必然选择，也成为海洋生态保护及渔业管理领域重要的研究主题之一。

一、海洋保护区界定

（一）海洋保护区定义

海洋保护区是作为一种经济有效的渔业资源维护、生物多样性保全以及满足其他人类保护目的的管理工具被提出来的（PDT, 1990; Bohnsack, 1996; Allison et al, 1998）。目前，海洋保护区已经成为一种受欢迎的、预防性的海洋保护及渔业管理工具，并在世界范围内不同的自然地理、生态环境、社会制度和文化及政治背景下得到了广泛发展。但在学术界及管理领域，有关海洋保护区的确切概念界定并没有达成一致意见，在世界各地存在多种不同的海洋保护区定义方法。按照世界自然保护联盟（IUCN）推荐的定义，海洋保护区是"任何通过法律程序或其他有效方式建立的，对其中部分或全部环境进行封闭式保护的潮间带或潮下带陆架区域，以及其上覆水体和相关的动植物群落、历史及文化属性"（IUCN, 1994）。可见海洋保护区不仅具有生物及生态学属性，同样具有历史文化属性以及相关联的经济学属性。其中，海洋生态保护属性与社会经济属性之间的冲突正是海洋保护区研究与实践所面临的最大挑战。

（二）海洋保护区的经济目的

根据基本的保护理论，海洋保护区的建立一般具有生态与社会经济双重目标。生态目标在于恢复遭受破坏的栖息地，维持生物多样性，保护海洋生物以及提供海洋科学研究试验场所；而社会经济目标则是恢复商业及休闲渔业，创造生态旅游体验，提供经济发展机遇等（Christie et al, 2003）。按照不同的保护对象和保护目标，基于国家自然保护联盟的保护区分类标准（IUCN,

1994），海洋保护区可划分为六大类，即严格的海洋保护区、海洋国家公园、海洋自然纪念地、海洋生境/物种保护区、海洋景观保护区和海洋资源管理区。不同类型的海洋保护区建立目的存在明显的差异，有些是纯粹的海洋自然生态保护区，不具备经济开发价值，如严格的海洋保护区；有些是保护型的资源开发区，可以进行适当的渔业、游憩及相关产业开发活动，具有很高的经济收益，如海洋国家公园、海洋自然纪念地和海洋景观保护区；也有些是为了物种保护及自然资源可持续利用，具有很高的潜在经济效益，如海洋生境/物种保护区和海洋资源管理区（见表1）。在很多情况下，经济效益的大小是决定海洋保护区选择与管理成败的关键因素。

表1　　　　　　　　海洋保护区类型与管理目标

管理目的 \ 类型	严格的海洋保护区	海洋国家公园	海洋自然纪念地	海洋生境/物种保护区	海洋景观保护区	海洋资源管理区
科学研究	1	2	2	2	2	3
荒野地保护	2	—	3	3	—	2
物种及基因多样性保护	1	1	1	1	2	1
环境维持	2	1	—	1	2	1
特定自然及文化属性保护	—	2	1	3	1	3
游憩和娱乐	—	1	1	3	1	3
教育	—	2	2	2	2	3
自然资源可持续发展	—	3	—	2	2	1
文化及传统属性维护	—	—	—	—	1	2

说明："1" = 主要目的；"2" = 次要目的；"3" = 潜在利用目的；"—" = 不适用。

二、海洋保护区的经济学属性

（一）海洋保护区的经济价值

近年来，随着人们海洋保护区意识的提高，海洋保护区概念得到了快速普及，有关海洋保护区的价值也得到人们越来越多的认可。如海洋保护区可以提高保护海域的生物量，增加生物多样性并恢复当地遭到破坏的生态系统等，这些变化将为海域利用者提供直接或间接的生态服务价值。海洋保护区可以提供的生态服务价值类型多样，其中一些属于市场价值，可以通过市场交易来反映，如渔获物；而另外一些则属于非市场价值，不能通过市场价格来反映其经济价值。在很多情况下，海洋保护区的价值被明显低估，如大型综合性海洋保护区的生物多样性保护价值和渔业增强价值。从整体上来看，海洋保护区至少可以提供以下四种类型的经济价值（NRC，2001）：

（1）与消耗性开发活动有关的市场价值，如海洋保护区周边开放海域渔获物的增加及渔业产量的稳定。这类价值很容易确定，并且可以量化。

（2）非消耗性的体验及娱乐价值，即公众欣赏海洋保护区内海洋生物及非生物景观的支付意愿，如潜水者赋予多样性高、物种丰富的海洋生物栖息地很高的主观价值，这种价值因海洋保护区的建立而提升。

（3）存在价值，指人们赋予海洋保护区内海洋生物多样性或独特的海洋生态系统的存在以货币价值，而不论其是否从海洋保护区获得直接或间接市场化价值，如海洋保护区的科研及教育价值。

（4）遗产价值。为子孙后代保留一个功能与结构相对完整的代表性海洋保护区系统，使未来的人类有一个公平的海洋开发利用机会。

（二）海洋保护区建设的经济动机

所有的环境保护问题都涉及到科学、社会与经济因素的影响。一般观点认为无论一个保护区的设计在科学上是多么合理，最终决定其成败的关键是社会经济因素，而成本最小化是保护区设计的核心问题（Stewart & Possingham，2005）。在现实生活中，经济因素在传统的海洋保护区决策过程中并没有得到充分体现，多数海洋保护区的建立通常只考虑到生物及生态效益，或者是面临社会压力，特别是环境保护团体游说的结果。在海洋保护区的日常管理中，特别是在解决不同用户冲突以及获取公众对海洋保护区管理的支持时，经济因素就成为一个重要的砝码（Sumaila et al，2000），而且短期的社会和经济成本经常成为海洋保护区规划与实施的障碍（Beattie et al.，2002），因此有必要将社会经济因素纳入到海洋保护区管理决策中。

在海洋保护区建设过程中，经济动机是冲突产生的根源，也是解决冲突的切入点。考虑到海洋保护区建设给现有资源利用者带来的短期经济损失，海洋保护区建设通常首先选择那些机会成本低的区域。这些区域可能已经被过度开发，或者经济利用价值不高，如果采取封闭管理措施对现有资源利用者不会造成很大的经济损失。但这种选择使海洋保护区的保护效益和渔业价值大打折扣，因为很多保护效益只有在未进行过度开发的种群中才能发挥出来，如产卵种群的维持、种群生存、栖息地保护以及灾后种群的快速恢复等。

由于海洋保护区的长期效益是分散的，尽管总体来看在经济上是有利的，但短期损失却集中在渔业等少数特定利益集团身上。对渔业而言，海洋保护区所带来的是短期经济利益的损失，这与渔民现有的短期利益最大化的动机相抵触。渔民不愿意在传统的渔业管理基础上同时考虑海洋保护区，可能是害怕对其捕捞活动的双重控制。海洋保护区建设意味着渔民不仅要放弃一些传统的渔场，还要被迫在其他渔场上接受捕捞强度控制，其直接经济成本是显而易见的，而长期的保护效益却并不确定（Sumaila et al, 2000）。除了渔业外，其他海洋开发活动也受到海洋保护区的影响，如滨海旅游、海上娱乐等，但这些产业开发活动却是海洋保护区建设的主要受益者，也是海洋保护区长远经济价值的主要表现领域。长期来看，海洋保护区的游憩及相关价值往往要高于其渔业价值（Hoagland et al, 1995），这也是很多海洋保护区得以建立与成功维护的基础。

依据传统的保护经济学观点，人们对海洋保护区的接受程度取决于海洋保护区给各相关利益方所造成的不同影响，成本效益分析最终决定了海洋保护区的选择。海洋保护区的受益者既包括看重海域自然价值与原始海洋生态系统，希望保留海洋自然景观并欣赏海洋美景的公众与游客，也包括那些希望获得长期稳定渔业产量的渔民。在现有的海洋管理体制下，各种海洋相关法律、法规已经确定了海洋生态系统的价值分配框架。任何涉及海洋保护区的政策变化都会改变现有的利益分配格局，使不同的利益集团之间出现新的利益冲突，并最终达成新的均衡。一般情况下，从海洋保护中得到的排他性收益越大，其参与制定新的保护性管理政策的积极性也就越高，对新的保护政策的制定影响也就越大。但有些海洋保护受益方的利益在政策制定过程中却得不到充分体现，这通常属于公共产品的受益者，由于这类效益对任何人而言都不是专有的，不可能通过市场手段来进行分配。因此海洋保护区建设的价值链更多地体现在一些强势集团的利益基础上，如渔业、旅游业等，而众多公众的利益却并没有得到充分反映。资源开发权，或者说是经济动机有可能成为决定海洋保护区是否作为一种海洋资源管理工具选择的决定因素（NRC，2001）。

（三）海洋保护区的成本效益分析

海洋保护区的经济效益一般体现在两个方面：一是海洋保护区建立后所形成的非消耗性海洋资源开发机会，如海上游憩开发效益；二是通过维持现有海洋经济生物种群的持续生存能力来保证未来的渔业效益。对于不同的用户而言，海洋保护区建设所带来的成本效益是不同的。海洋保护区在增加非消耗性海洋资源开发机会，并产生相应的经济效益，如海上游憩效益的同时，也可能迫使一些行业放弃开发机会，产生明显的机会成本，如海洋油气及矿产开发等，但从长远来看，海洋保护区可以使大部分用户受益。如针对渔业开发来说，海洋保护区的影响存在很大差异。对于已经过度捕捞的渔业种群而言，海洋保护区具有恢复衰退渔业种群的潜在效益；而对于那些资源丰富、没有过度开发的渔业而言，海洋保护区的长期渔业效益是不明显的。总的来看，在一定条件下，海洋保护区对于任何类型的渔业都是一种有益的辅助管理工具，具有长期的经济可行性。

从成本效益分析的角度来看，海洋保护区的成本效益评估既包括市场价值的评估，也包括非市场价值的评估。如果只进行市场条件下的成本效益评估，很多海洋保护区的建立并不能提供净经济效益，如海洋保护区的保护利用活动具有很高的社会与环境价值，但包括渔业及旅游业在内的经济价值却很小，因此在任何有关海洋保护区的成本效益分析中，包括对非市场价值的评估是非常重要的。如果海洋保护区所产生的长期可持续渔业收益及其他经济收益大于海洋保护区建设本身的投入及放弃的渔获量等短期成本，可以最终证明海洋保护区优于传统的渔业管理工具（见表2）。

表2　　　　　　　海洋渔业保护区的潜在成本—效益分析

项目	成　本	效　益
产量	● 捕获量减少（至少是临时性的） ● 迁移性鱼群效益的不确定性 ● 严重过捕的种群，大面积的禁捕区可能对其他渔业产生不良影响	● 长期的资源稳定性 ● 地方种群繁殖力的增加 ● 未来产量和补充量的增加 ● 副渔获物的减少
渔场变动	● 给当地渔民及渔业相关产业造成困难，如到更远的渔场捕鱼 ● 增加其他开放海域的捕捞压力	● 减少对过捕种群的开发 ● 对基本渔业生境的保护
执行	● 海洋保护区边界管理，执行成本随保护区面积的增加而增高	● 执行更加有效。特别是当渔民理解并服从保护区管理时，其效果更明显
管理	● 对监测与研究更高的要求，如更多的种群生命史、扩散类型、捕捞力变化及生境的数据	● 对统计参数估计的改善，如种群自然死亡率
经济活动	● 海洋保护区给周边社区带来不均衡的影响 ● 短期利润的减少	● 游憩机会的潜力更大 ● 非消耗性用户的潜在冲突减少 ● 减少因自然波动或过捕造成的种群崩溃
非市场价值	● 传统的渔场和入渔权的丧失	● 欣赏更加原始自然的海洋生态系统 ● 对景观生境和珍稀物种的保护

资料来源：美国国家研究委员会，《海洋保护区白皮书（2001）》。

三、海洋保护区经济学研究进展

在过去的二十多年间,世界海洋保护区建设取得突破性进展,海洋保护区数量快速增长,有关海洋保护区的研究也急剧增长,出现了大量有关海洋保护区的生物学实证(如 Russ & Alcala, 1994, 1996a/b; Roberts & Polunin, 1994; Roberts, 1995a/b; Murawski et al, 2000; Halpern, 2003; Alcala et al, 2005)和理论研究(如 Polacheck, 1990; Quinn et al, 1993; DeMartini, 1993; Man et al., 1995; Sladek Nowlis & Roberts, 1999; Guénette & Pitcher, 1999; Mangel, 2000a; Gerber et al, 2002),为海洋保护区的生态学价值提供了强有力的证据。但与此同时,有关海洋保护区的经济学研究却很少,且多数属于理论研究,部分原因在于经济学实证研究的困难。在进行深入的经济学实证分析之前,政策制定者需要对海洋保护区的生物及经济成本效益作出预测。这面临两方面困难:一是决定产出的内在生物经济学驱动因素并不明确,具体取决于特定的渔业类型;二是目前针对海洋保护区的研究很少对其影响的评估结果进行描述,或对不同模型之间的假设条件进行比较(Sanchirico, 2002)。

对于目前的海洋保护区经济学研究,苏迈拉等(Sumaila et al, 1999)认为主要存在两个方向:一是决定海洋保护区净经济价值的成本效益分析,除了传统的海洋保护区渔业价值外,主要考虑非消耗性开发活动,如游憩机会增加的可能性。这种价值主要通过条件价值法、享乐定价法和旅行成本法等常见的非市场价值方法来进行评估(Dixon, 1993; Sobel, 1993);二是生物经济学模型(Holland & Brazee, 1996; Hannesson, 1998; Sumaila, 1998; Pezzey et al, 2000; Sanchirico & Wilen, 1999a/b)。前者主要基于实证分析,而后者则主要突出生物经济学理论模型的应用。

(一)海洋保护区实证经济学研究

1. 市场价值研究

由于现代意义上的海洋保护区研究历史并不长,且主要集中在海洋保护区的生物学效益研究上。而有关海洋保护区成本效益的实证经济学研究很少,已有的少量有关海洋保护区的实证经济学研究也主要集中在渔业及游憩经济效益方面。现有的渔业证据显示有很多海洋保护区的确可以在保全生物多样性的同时,提高当地的渔业产量(如 Alcala et al, 2005; Rodwell & Roberts, 2004;

Agardy, 2001；Jennings & Karser, 1998），包括贝类（Wallace, 1999；Murawski et al, 2000）、甲壳类（Kelly et al, 2000）、海鞘类（Castilla, 1999）及各种鱼类等（Willis et al, 2001；Wantiez et al, 1997；Holland & Brazee, 1996；Murawski et al, 2000），从而提升当地的渔业经济效益。但也有一些研究得出不同的结论，认为海洋保护区对外部渔业种群的增强作用并不明显（Hasting & Botsford, 1999；Sumaila et al, 2000），海洋保护区对当地捕捞产量没有影响或影响很小（Willis et al, 2003；Gerber et al, 2002；Horwood et al, 1998）。此外，海洋保护区的建立还以牺牲部分渔场及短期渔业产量为代价，而其长期渔业效益却取决于保护区内目标物种的移动性、生命史特征、现有捕捞水平及捕捞强度对海洋保护区建立的反应等多种因素，存在很高的不确定性。

在游憩价值方面，研究者发现很多海洋保护区对游客产生的吸引力及相关经济效益远远超过其渔业效益，如在岩礁类海洋保护区中，与岩礁海域有关的鱼类大小、丰度和多样性的增加对于潜水者来说比岩礁本身更具价值（Williams & Polunin, 2000），很多加勒比海的潜水者更多地选择在海洋保护区内潜水，游憩业收入已经成为当地海洋保护区主要的经济收入来源。如1956年建立的维京群岛国家公园，20世纪80年代早期的年均运营成本大约是210万美元，而来自旅游及娱乐的收益为2 330万美元，其中直接收益330万美元，间接收益2 000万美元，是保护区运营成本的10倍多。1987年建立的荷属安的列斯萨巴岛海洋公园，1992年上半年吸引的潜水者只有9 200人次，但到了1994年就达到3万人次；而同属安的列斯群岛的博奈尔海洋公园仅1991年吸引的潜水者就达到17万人次（Dixon, 1993），可见当地海洋保护区内游憩业发展的潜在市场价值远高于当地渔业。

在海洋保护区建设的直接成本方面，根据Balmford等（2004）的测算，一个覆盖全球海域20%~30%的海洋保护区网络的总运营成本在50亿~190亿美元/年，这就需要在现有的海洋保护支出方面大约提高两个数量级。但这种投入的收益也是显著的，除了直接的渔业产量增加外，海洋保护区所维持的非消耗性海洋服务价值将高达4.5万亿~6.7万亿美元。依据不同的保护水平，满足最低安全标准的海洋保护区系统的成本收益率保守估计为1∶100，即1美元的海洋保护区投入可以产生100美元的收益，海洋保护区的经济效益还是相当可观的，但目前世界海洋保护区建设投入的不足严重制约了海洋保护区经济效益的实现（Balmford et al, 2002）。

2. 非市场价值研究

海洋保护区除了具有渔业与游憩等市场价值外，还具有海洋生物多样性保护、生境保全以及其他生态服务功能等多种价值。其中只有部分渔业及游憩价值可以通过水产品及旅游市场价值来衡量，而其他多种生态系统服务功能都难以通过市场价值来衡量，只能通过一些非市场价值评估手段来完成。这些评估方法包括揭示性偏好方法（revealed preference approaches）和陈述性偏好方法（stated preference approaches）两大类，其中揭示性偏好方法是指利用消费者实际消费行为来推断相关非市场化产品的支付意愿，如旅行成本法、享乐定价法、随机效用模型等；而陈述性偏好方法则是指通过问卷调查直接获取消费者对非市场产品或服务的主观支付意愿，如条件价值评价法（见表3）。

表3　　　　　　　　　　　　非市场价值评估方法

	实际行为（揭示偏好）	假设市场（陈述偏好）
直接	直接观察 市场价格法 模拟市场法（实验经济学）	直接假设 投标法（bidding games） 条件价值评估法（contingent valuation）
间接	间接观察 生产函数模型（production function models） 替代支出法（surrogate expenditures） 随机效用模型（random utility） 旅行成本法（travel cost） 享乐定价法（hedonic pricing） 规避支出法（avoidance expenditures） 公民投票法（teferendum voting）	间接假设 条件排序法（contingent ranking） 条件行为法（contingent activity） 条件公决法（contingent referendum）

资料来源：Freeman A M. 1993. The measurement of environmental and resource values. Resources for the Future, Washington DC, USA.

利用非市场价值评估方法，可以实现对海洋保护区多种游憩价值及非使用价值的量化评价。霍格兰等（Hoagland et al，1995）对过去15年间的有关海洋保护区非市场价值的各类研究进行了回顾，发现了多项利用不同的非市场价值评估方法对海洋保护区游憩价值进行评估的研究（见表4）。但由于学术界在非市场价值研究方法上还存在较大争议，在很多情况下其研究结论并不被传统经济学界所接受。

表4　　　　　　　　　　　　海洋保护区非市场价值评估

海　域	地点	时间（年）	平均价值[3] （美元/人/天）	模型	平均年限（年）	多目的地
约翰潘纳坎谱珊瑚州立公园	美国佛罗里达州	1988~1989	356~533	TCM	47	是
加拉帕戈斯国家公园	加拉帕戈斯群岛	1986	439	HA	53	否
大堡礁海洋公园	澳大利亚北昆士兰	1985~1986	228[d] 138[i]	TCM	—	否

续表

海域	地点	时间（年）	平均价值[3]（美元/人/天）	模型	平均年限（年）	多目的地
博内尔海洋公园[1]	荷属安的列斯	1991	132	—	—	否
威尔福利特港[2]	美国马萨诸塞州	1994	66[d] 87～111[i]	CVM	56[d] 45[i]	否[d] 是[i]

说明：1 表示每次潜水的平均经济影响价值；2 表示永久保护的年度支付意愿；3 表示 1995 年美元；d 表示国内或当地居民；i 表示国外或旅游者；TCM 表示旅行成本法；HA 表示享乐价值法；CVM 表示条件价值法。

（二）海洋保护区理论经济学研究

经济效益分析是海洋保护区理论研究的重要内容之一，在传统的种群生态学理论研究中纳入经济学因素所形成的生物经济学模型已经成为海洋保护区理论模型研究的重要分支，经济学与生态学理论的结合将更有利于理解和推动海洋保护区的发展。生物经济学模型是种群动态变化模型与经济学成本效益优化模型的有机结合，其优化策略既要考虑生物学产量的最大化，也要考虑经济收益的最大化，是建立在生物学模型基础上的经济收入最大化模型。生物经济学模型除了需要考虑一些种群变化的关键生物学过程，如海洋保护区与捕捞区之间的个体迁移率和物质通量外，还要考虑海洋保护区对当地渔民行为的影响以及自然环境变化对海洋保护区和被捕捞种群的影响（Sanchirico et al, 2003）。此外，经济学分析不仅需要自然资源生物经济学模型中传统的优化和行为方法，也需要多目标分析、条件价值评估以及社会经济研究来提供相关决策信息（Sumaila & Charles, 2002）。

现有的生物经济学模型形式多样，假设条件、研究对象及研究内容也存在很大差异，这在很多情况下造成研究结果的不可比性。迄今为止，大多数生物经济学模型研究针对单物种种群进行，其中有些考虑到种群空间分布因素的影响（如 Holland, 1998；Sanchirico & Wilen, 1999a），另一些则不考虑空间因素（如 Holland & Brazee, 1996；Hannesson, 1998），只有极少数生物经济学研究基于多物种或生态系统模型并进行了空间（Walters, 2000；Pitcher et al, 2000）及非空间（Sumaila, 1998）分析。但无论是哪一类模型，都只是针对海洋保护区管理单方面进行，没有考虑到渔民等其他方面的反映，只有苏迈拉（Sumaila, 2002）尝试了同时对海洋保护区及渔业双方进行双边分析，但结果并不理想。

1. 集合种群模型

集合种群（metapopulation）是一组通过个体迁移相互联接的地方种群的

集合，这些地方种群通常生活在隔离的生境斑块中，其隔离程度取决于斑块之间的距离。而集合种群模型则是将这些地方种群看做是一个整体，并将幼体扩散与成体迁移结合在一起进行综合分析的生物经济学模型。除了考虑捕捞所带来的灭绝效应外，近年来出现的有关海洋保护区的集合种群模型研究并不多，而且多结合经济学的成本效益分析进行。如考虑幼体扩散（Brown & Roughgarden, 1997；Pezzey et al, 2000；Tuck & Possingham, 2000）和成体迁移（Sanchirico & Wilen, 1999a, 2001, 2002）的集合种群生物经济学模型, 也有少数几项研究将成体溢出和幼体扩散整合在一个集合种群生物经济学模型中进行分析（Holland & Brazee, 1996；Rodwell et al, 2001）。

集合种群模型并不具备空间结构性，斑块分布的空间类型可以是任意形式的，定居行为在所有斑块间均等发生。曼（Man, 1995）等利用一个多斑块集合种群模型来研究海洋保护区对珊瑚礁种群的保护及渔业开发效益，结果显示海洋保护区的建立对于集合种群的可持续开发性具有显著效益，海洋保护区可以通过为过度捕捞的斑块提供种群补充源来弥补渔业开发的影响，以预防地方性灭绝现象。当所有斑块的一半左右被补充个体所占据时，其产量和复合种群丰度达到最大；如果达不到，则需要增加海洋保护区的面积。在确定地点，总的渔业捕获量取决于潜在的生境异质性，不仅仅受到海洋学及生物学条件的影响，也包括空间经济特征。如某地的渔业主要以当地的幼鱼生产为主，定居与成体生物量扩散水平不高，那么就需要考虑关闭成本低的斑块；而对于没有成体生物量溢出，但具有幼体扩散效应的渔业，可能就要选择成本高的斑块。

2. 空间经济学模型

空间经济学模型主要考虑种群空间分布结构对海洋保护区经济效益的影响。由于海洋环境的空间异质性以及自然生境的丧失与碎片化，生境空间结构变化对于目标种群的生存与发展具有越来越重要的意义，因此应该将目标种群栖息地的空间结构纳入生物经济学模型以帮助理解幼体扩散、成体迁移以及对不同年龄栖息地需求的影响。阿特伍德和班尼特（1995）利用一个简单的单物种空间结构模型来比较三种具有不同生活史（寿命、繁殖和迁徙）的物种，研究结果显示迁徙影响到重建种群所必须的海洋保护区大小。如果海洋保护区的目的在于限制捕捞死亡率并比较不同的管理效益，种群分布的空间动力学及捕捞强度也应包括在内（Sumaila et al, 2000）。由于没有考虑到捕捞活动分布的空间异质性，单位补充产量模型系统地夸大了处在过度捕捞渔业海域的海洋保护区的渔业增强效益；而在开发水平很低的渔业中，对于捕捞强度空间分布异质性的忽略则夸大了海洋保护区建设所带来的渔获量损失（Smith, 2004）。具有明确空间分析的海洋保护区模型可分为两类：一类是利用两种斑块，即保护区和捕捞区。这类模型只提供有限的空间分析，主要用于以成体溢出为主要个体输出来源的海洋保护区研究，特别适用于非定居性成体及位于热带珊瑚礁生态系统中的小型海洋保护区研究；另外一类模型则是利用多个斑块或斑块之间具有连续的空间结构。这类模型针对大尺度的空间迁移问题，经常面对幼体具有高扩散潜力的物种种群变化。

目前，基于空间结构的海洋保护区生物经济学模型比较少见，但这类研究趋向于多物种分析，甚至考虑整个小生态系统，研究难度较大，但更贴近实际。圣基里和威伦（Sanchirico & Wilen，1999a）指出海洋保护区对捕捞的影响在很大程度上取决于渔民如何重新配置其捕捞力，渔民可能将其捕捞力转移到租金相对较高的海域。这种现象在空间上造成了捕捞的"经济梯度"，可能与通过幼体输出及成体迁移所产生的"生物梯度"存在很大的差异。在很多情况下，渔业可以从收益率较低的海洋保护区获利，而并非那些建立在具有独特生物学特征海域中的海洋保护区。霍兰德（Holland，1998）在一个具有空间结构的、多区域、多物种模型中增加了渔民对渔场的选择，结果发现当捕捞强度处在最适水平时，海洋保护区的渔业效益并不显著，但产卵种群生物量可以维持在很高水平。海洋保护区对不同渔民所产生的影响是不同的，有些可能受益，但有些也可能受损。在美国西海岸加利福尼亚州海胆渔业中，史密斯和威伦（Smith & Wilen，2003）运用空间经济学模型发现如果渔民捕捞力的空间分布行为是理性的，在过度捕捞的海域建立海洋保护区将减少贴现租金；但如果不是的话，则提高贴现租金。空间经济学模型研究说明海洋保护区选址需要生物学与经济学的双重考虑。

3. 确定性模型

确定性模型（deterministic model）是在确定的前提条件下，对海洋保护区内的受保护种群进行生物学及经济学模拟的方法。霍兰德与布拉兹（Holland & Brazee，1996）的模型是最早研究海洋保护区渔业影响的经济学模型之一，他们利用一个确定的模型发现海洋保护区的相对效益取决于保护区对捕捞海域渔获量的影响以及贴现率水平。海洋保护区所产生的短期渔业损失越大，贴现率越高，其渔业效益也就越小。同时他们还发现在很高的捕捞强度下，海洋保护区具有保险作用，如果捕捞强度得到很好的控制，海洋保护区可能没有任何价值。在随后的研究中，霍兰德（2000）发现如果总渔获量或捕捞强度可以被有效控制，在确定条件下海洋保护区是多余的；但如果存在过度捕捞现象，海洋保护区则具有一定的经济效益。这一观点也得到安德森（Anderson，2002）和汉内松（Hannesson，1998）研究结果的支持。

佩兹等（Pezzey et al，2000）、圣基里科和威伦（Sanchirico & Wilen，

2001）的研究也都证明在密度依赖条件下，海洋保护区可以增加被捕捞种群的丰度。当种群处在过度捕捞状态时，有时甚至可以提高捕捞海域的总产量。佩兹等（Pezzey et al, 2000）还发现在海洋保护区建立之前，如果目标种群数量达不到其海域承载力水平的50%时，海洋保护区能增加其均衡产量。在种群丰度低的海域建立海洋保护区，其禁捕措施所产生的边际效益将超过其成本（Grafton et al, 2004）。在确定性模型研究中，海洋保护区的渔业效益经常被低估。即使在捕捞强度适中、无不良风险、种群持续生存并且没有前期过度捕捞的条件下，海洋保护区也具有经济价值。对于很多已开发渔业种群，即使捕捞产量处在最适状态，早期建立的小型海洋保护区也可以创造经济效益（Grafton et al, 2003）。

4. 不确定性模型与随机模型

由于没有考虑捕捞种群内在的不确定性，以及海洋保护区在面对意外冲击和管理失败时所具有的缓冲作用，确定性模型可能低估了严格的海洋保护区价值（Gerber et al, 2003）。进一步而言，没有明确地考虑到渔业冲击的模型难以决定海洋保护区的适当大小与设立地点。海洋保护区可以作为一种抵御不确定性的屏障，特别是那些与捕捞产量密切相关的不确定性（Ludwig et al, 1993；Botsford et al, 1997）。在不确定性条件下，劳克（Lauck, 1998）等认为目标种群大小存在管理上的不确定性，海洋保护区作为一种预防性措施，其面积应该增加以确保捕捞种群的持续性。在确保资源长期稳定性的前提下，甚至可以实现更高程度的开发。苏迈拉（Sumaila, 1998）和曼格尔（Mangel, 1998, 2000b）也发现海洋保护区大小和不确定性程度之间存在负相关关系。若要维持更高的产量水平，就需要更大面积的海洋保护区，这与加尔特和皮切尔（Guénette & Pitcher, 1999）的研究结果类似。他们认为海洋保护区通过维持高水平的产卵生物量和提升高开发强度下的补充成功率就可以增加种群持续性。

多耶思和贝奈斯（Doyen & Bene, 2003）发现不确定性越高，所需要的海洋保护区面积也就越大，从而使最小生存种群得以维持。在某种确定的条件下，海洋保护区有助于保持最小生存种群，并提高捕捞海域稳定的渔业产量。除了海洋保护区的"长期效益"外，康拉德（Conrad, 1999）发现在环境条件不变时，如果捕捞强度控制处在优化状态时，海洋保护区也可以减少渔业经济收益；而与之对照的是在变动环境条件下，海洋保护区可以通过减少种群数量的波动来获利。史莱德志和罗伯茨（Sladek Nowlis & Roberts, 1999）、曼格尔（Mangel, 2000b）以及汉内森（Hanneson, 2002）的研究也发现了类似的结果。最近的海洋保护区研究在生物经济学模型中整合了两种不确定性，即环境随机变化与负面影响。在宽泛的参数值范围内，格拉夫顿（Grafton, 2004）

等发现只要海洋保护区面积大于零,即使其面积很小,在面对意外冲击时也比没有保护区时获得更多的捕捞收益,这与前面确定性模型的结论有些矛盾。

四、结语

从公共管理的角度来看,海洋保护区建设属于一种公共投资,其决策过程理应建立在成本效益分析基础之上。由于海洋保护区发展在产生生态效益的同时,会给当地渔民及其他相关利益集团造成一定的经济损失。为了获取当地社区及渔民的支持,海洋保护区发展在关注生态环境问题的同时也要兼顾社会经济问题,即海洋保护区的设计与规划既需要考虑海洋生物多样性保护、海洋生态系统结构与功能的维持,也要考虑其社会经济效益。在海洋保护区建设博弈中,管理者的意图在于限制海洋资源开发活动,维持特定海域生态系统的完整性及其生态服务功能;渔民的目的是将海洋保护区建设所带来的损失最小化,促使海洋保护区建设远离资源丰富的海域;而旅游开发者及潜水者则要求保护具有更高游憩价值的海域,这种不同利益集团之间的冲突使海洋保护区的发展受到多种社会经济因素,包括贴现率、资源租金以及管理成本等因素的制约。

迄今为止,海洋保护区建设的主要推动者还是环境保护团体及生态经济学者,多数海洋保护区的目的也是保护脆弱的海洋生态系统、濒危物种及特定生境等,渔业及其他产业发展目的并不多见。传统的海洋保护区设计与规划主要依赖特定海域或生态系统的生物学及海洋学等自然科学信息,经济因素一般在海洋保护区规划中并没有被考虑,或者说社会经济问题很少能与生态环境问题得到同等重视,这就造成海洋保护区在选择设计过程中存在一定的社会政治偏差。一个成功的海洋保护区的设计、创建与管理过程不仅需要生物学信息,也需要广泛的社会经济信息。对于管理者及广大公众而言,海洋保护区的社会经济价值与生态学价值同样重要,有时甚至是海洋保护区建设得以顺利实施的关键因素。鉴于海洋保护区在海洋生态系统保护、生态服务功能维持方面的重要作用,努力争取公众及产业部门的支持,推动世界范围内海洋保护区网络的发展是沿海地区海洋管理部门所面临的最大挑战。海洋保护区建设的经济有效性,或者说是成本效益率的高低是决定海洋保护区获取公共资金投入及公众支持的基础。因此建立在成本效益分析基础上的经济学研究对于提高海洋保护区的社会支持度及管理成效,成功实现海洋保护区的生态及社会经济目标具有重要的现实意义。

2000年底,美国已经推出了海洋保护区社会科学研究战略,将社会科学研究正式纳入海洋保护区的规划、管理与评估过程。其中经济学研究是最重要的主题之一，主要包括与海洋保护区建设相关的各种市场与非市场价值、成本效益分析及造成的不同经济影响等内容，为海洋保护区的经济学研究奠定了制度基础。经济学已经成为世界各地海洋保护区理论与实践领域不可或缺的重点研究内容之一，对于海洋保护区理论的长远发展具有深远的影响。

参考文献

［1］Agardy T. 2001. An environmentalist's perspective on responsible fisheries：*the need for holistic approaches*. In：M Sinclair, G Valdimarsson, et al. 2003. Responsible Fisheries in the Marine Ecosystem. Minister of Fisheries/FAO, October 2001. Reykjavik, Iceland, pp. 65 – 85.

［2］Alcala A C, G R Russ, A P Maypa, et al. 2005. A long-term spatially replicated experimental test of the effect of marine reserves on local fish yields. Canadian Journal of Fisheries and Aquatic Sciences, 62：pp. 98 – 108.

［3］Allison G W, J Lubchenco, and M H Carr. 1998. Marine reserves are necessary but not sufficient for marine conservation. Ecological Applications, 8（1）(Supp)：pp. S79 – S92.

［4］Anderson L G. 2002. A bioeconomic analysis of marine reserves. Natural Resource Modeling, 15（3）：pp. 311 – 334.

［5］Attwood C G and B A Bennett. 1995. Modeling the effect of marine reserves on the recreational shore-fishery of the South-Western Cape, South Africa. South African Journal of Marine Science, 16：pp. 227 – 240.

［6］Balmford, A, P Gravestock, N Hockley, et al. 2004. The worldwide costs of marine protected areas. Proceedings of the National Academy of Sciences, Early Edition：pp. 1 – 4.

［7］Balmford A, A Bruner, P Cooper, et al. 2002. Economic reasons for conserving wild nature. Science, 297：pp. 950 – 953.

［8］Beattie A, U R Sumaila, V Christensen, et al. 2002. A model for the bioeconomic evaluation of marine protected area size and placement in the North Sea. Natural Resource Modeling, 15（4）：pp. 414 – 437.

［9］Bohnsack J A. 1996. Marine reserves, zoning, and the future of fishery management. Fisheries, 21（9）：pp. 14 – 16.

［10］Botsford L W, J C Castilla, and C H Peterson. 1997. The management of fisheries and marine ecosystems. Science, 277：pp. 509 – 515.

[11] Brown G and J Roughgarden. 1997. A metapopulation model with private property and a common pool. Ecological Economics, 22: pp. 65 – 71.

[12] Castilla J C. 1999. Coastal marine communities: trends and perspectives from human exclusion experiments. Trends in Ecology and Evolution, 14: pp. 280 – 283.

[13] Christie, P, B J McCay, and M L Miller et al. 2003. Toward developing a complete understanding: a social science research agenda for marine protected areas. Fisheries, 28 (12): pp. 22 – 26.

[14] Conrad J M. 1999. The bioeconomics of marine sanctuaries. Journal of Bioeconomics, 1: pp. 205 – 217.

[15] DeMartini E E. 1993. Modeling the potential for fishery reserves for managing Pacific coral reef fishes. Fishery Bulletin, 91 (3): pp. 414 – 427.

[16] Dixon, J A, 1993. Economic benefits of marine protected areas. Oceanus, 36: pp. 35 – 40.

[17] Doyen L and C Béné. 2003. Sustainability of fisheries through marine reserves: a robust modeling analysis. Journal of Environmental Management, 69: pp. 1 – 13.

[18] Freeman A M. 1993. The measurement of environmental and resource values. Resources for the Future, Washington DC, USA.

[19] Gerber L R, L W Botsford, A Hastings, et al. 2003. Population models for marine reserve design: a retrospective and prospective synthesis. Ecological Applications, 13 (1) (supp): pp. S47 – S64.

[20] Gerber L R, P M Kareiva, and J Bascompte. 2002. The influence of life history attributes and fishing pressure on the efficacy of marine reserves. Biological Conservation, 106: pp. 11 – 18.

[21] Grafton, R. Q., T. Kompas, and V. Schneider, 2004. The bioeconomics of marine reserves: a selected review with policy implications. Economics discussion papers, No. 0405. University of Otago.

[22] Grafton, R. Q., P. V. Ha, and T. Kompas, 2003. On marine reserves: rents, resilience and 'rules of thumb'. Discussion paper 0308. Department of Economics, University of Otago, Dunedin, New Zealand.

[23] Guénette S and T J Pitcher. 1999. An age-structured model showing the benefits of marine reserves in controlling over-exploitation. Fisheries Research, 39: pp. 295 – 303.

[24] Halpern, B S, S Walbridge, K A Selkoe, et al. A global map of human impact on marine ecosystems. Science, 2008, 319 (5865): pp. 948 – 952.

[25] Halpern B S. 2003. The impact of marine reserves: do reserves work and does reserve size matter? Ecological Applications, 13 (1) (supp): pp. S117 – S137. Halpern B S. 2003. The impact of marine reserves: do reserves work and does reserve size matter? Ecological Applications, 13 (1) (supp): pp. S117 – S137.

[26] Hannesson R. 2002. The economics of marine reserves. Natural Resource Modeling, 15 (3): pp. 273 – 290.

[27] Hannesson R. 1998. Marine reserves: what would they accomplish? Marine Resource Economics, 13: pp. 159 – 170.

[28] Hastings A and L W Botsford. 1999. Equivalence in yield from marine reserves and traditional fisheries management. Science, 284: pp. 1537 – 1538.

[29] Hoagland P, Y Kaoru, and J M Broadus. 1995. A methodological review of net benefit evaluation for marine reserves. Environmental Department Paper No 27, The World Bank, Washington, DC.

[30] Holland D S. 2000. A bioeconomic model of marine sanctuaries on Georges Bank. Canadian Journal of Fisheries and Aquatic Sciences, 57 (6): pp. 1307 – 1319.

[31] Holland D S. 1998. The use of year around closed areas for the management of New England trawl fisheries, PhD Thesis, University of Rhode Island, Kingston, RI, USA.

[32] Holland D S and R J Brazee. 1996. Marine reserves for fisheries management. Marine Resource Economics, 11: pp. 157 – 171.

[33] Horwood J W, J H Nichols, and S Milligan. 1998. Evaluation of closed areas for fish stock conservation. Journal of Applied Ecology, 35: pp. 893 – 903.

[34] IUCN. 1994. Guidelines for protected area management categories. CNPPA with the Assistance of WCMC. IUCN, Gland, Switzerland and Cambridge, UK.

[35] Jennings S and M Kaiser. 1998. The effects of fishing on marine ecosystems. Advances in Marine Biology, 34: pp. 201 – 352.

[36] Kelly S, D Scott, A B MacDiarmid, et al. 2000. Spiny lobster, *Jasus edwardsii*, recovery in New Zealand marine reserves. Biological Conservation, 92 (3): pp. 359 – 369.

[37] Lauck T, C W Clark, M Mangel, et al. 1998. Implementing the precautiona-

ry principle in fisheries management through marine reserves. Ecological Applications, 8 (1) (supp): pp. S72 – S78.

[38] Ludwig D, R Hilborn, and C Walters. 1993. Uncertainty, resource exploitation, and conservation: Lessons from history. Science, 260 (2): p. 7, 36.

[39] Man A, R Law, and N V C Polunin. 1995. Role of marine reserves in recruitment to reef fisheries: a metapopulation model. Biological Conservation, 71 (2): pp. 197 – 204.

[40] Mangel M. 2000a. Trade-offs between fish habitat and fishing mortality and the role of reserves. Bulletin of Marine Science, 66: pp. 663 – 674.

[41] Mangel M. 2000b. Irreducible uncertainties, sustainable fisheries and marine reserves. Evolutionary Ecology Research, 2: pp. 547 – 557.

[42] Mangel M. 1998. No-take areas for sustainability of harvested species and a conservation invariant for marine reserves. Ecology Letters, 1: pp. 87 – 90.

[43] Murawski S A, R Brown, J J Lai, et al. 2000. Large-scale closed areas as a fishery management tool in temperate marine systems: The Georges Bank experience. Bulletin of Marine Science, 66 (3): pp. 775 – 798.

[44] NMPAC. 2003. Social science research strategy for marine protected areas. National Marine Protected Areas Center, MPA Science Institute, Santa Cruz, California, U. S.

[45] NRC. 2001. Marine protected areas: tools for sustaining ocean ecosystems. National Research Council, National Academy Press, Washington, DC.

[46] PDT. 1990. The potential of marine fishery reserves for reef fish management in the US southern Atlantic. Snapper-Grouper Plan Development Team Report for the South Atlantic Fisheries Management Council. NOAA Tech. Memo. NMFS-SEFC-261. NOAA, National Marine Fisheries Service, Southeast Fisheries Center, Miami, Florida.

[47] Pezzey J C V, C M Roberts, and B T Urdal. 2000. A simple bioeconomic model of a marine reserve. Ecological Economics, 33: pp. 77 – 91.

[48] Pitcher T J, R Watson, N Haggan, et al. 2000. Marine reserves and the restoration of fisheries and marine ecosystems in the South China Sea, Bulletin of Marine Science, 66: pp. 543 – 566.

[49] Polacheck T. 1990. Year around closed areas as a management tool. Natural Resource Modeling, 4 (3): pp. 327 – 354.

[50] Quinn J F, S R Wing, and L W Botsford. 1993. Harvest refugia in marine invertebrate fisheries: models and applications to the red sea urchin, Strongylocentrotus franciscanus. American Zoologist, 33: pp. 537 – 550.

[51] Richardson, E. A., M. J. Kaiser, G. Edwards-Jones, and H. P. Possingham, 2006. Sensitivity of marine-reserve design to the spatial resolution of socioeconomic data. Conservation Biology 20 (4): pp. 1191 – 1202.

[52] Roberts C M. 1995a. Effects of fishing on the ecosystem structure of coral reefs. Conservation Biology, 9 (5): pp. 988 – 995.

[53] Roberts C M. 1995b. Rapid build-up of fish biomass in a Caribbean marine reserve. Conservation Biology, 9 (4): pp. 815 – 826.

[54] Roberts C M and N C Polunin. 1994. Hol Chan: demonstrating that marine reserves can be remarkably effective. Coral Reefs, 13 (2): p. 90.

[55] Rodwell L D and C M Roberts. 2004. Fishing and the impact of marine reserves in a variable environment. Canadian Journal of Fisheries and Aquatic Sciences, 61: pp. 2053 – 2068.

[56] Rodwell L D, E B Barbier, C M Roberts, et al. 2001. A bioeconomic analysis of tropical marine reserve-fishery linkages: Mombasa Marine National Park. In: U Sumalia and J Alder (ed). Economics of marine protected areas. FCRR, 9 (8): pp. 183 – 197.

[57] Russ, G R and D C Zeller. 2003. From Mare Liberum to Mare Reservarum. Marine Policy, 27: pp. 75 – 78.

[58] Russ G R and A C Alcala. 1996a. Marine reserves: rates and patterns of recovery and decline of large predatory fish. Ecological Applications, 6: pp. 947 – 961.

[59] Russ G R and A C Alcala. 1996b. Do marine reserves export adult fish biomass? Evidence from Apo Island, central Philippines. Marine Ecology Progress Series, 132: pp. 1 – 9.

[60] Russ G R and A C Alcala. 1994. Sumilon island reserve: 20 years of hopes and frustrations. Naga, 17 (3): pp. 8 – 12.

[61] Sanchirico J N. 2002. Additivity properties in metapopulation models: implications for the assessment of marine reserves, Discussion Paper 02 – 66, Resources for the Future, Washington DC.

[62] Sanchirico J N, R Stoffle, K Broad, et al. 2003. Modeling marine protected areas. Science, 301: pp. 47 – 48.

[63] Sanchirico, J N, K A. Cochran, and P M, Emerson, 2002. Marine protected areas: economic and social implications. Discussion Paper 02 – 26. Resources for the Future. Washington, D. C.

[64] Sanchirico J N and J E Wilen. 2002. The impacts of marine reserves on limited-entry fisheries. Natural Resource Modeling, 15 (3): pp. 291 – 310.

[65] Sanchirico J N and J E Wilen. 2001. A bioeconomic model of marine reserve creation. Journal of Environmental Economics and Management, 42 (3): pp. 257 – 276.

[66] Sanchirico J N and J E Wilen. 1999a. Bioeconomics of spatial exploitation in a patchy environment. Journal of Environmental Economics and Management, 37 (2): pp. 129 – 150.

[67] Sanchirico J N and J E Wilen. 1999b. Marine reserves: is there a free lunch? Discussion Paper 99 – 09. Resource for the Future, Washington, DC.

[68] Sladek Nowlis J S and C M Roberts. 1999. Fisheries benefits and optimal design of marine reserves. Fishery Bulletin, 97 (3): pp. 604 – 616.

[69] Smith M D. 2004. Fishing yield, curvature and spatial behavior: implications for modeling marine reserves. Natural Resource Modeling, 17 (3): pp. 273 – 298.

[70] Smith M D and J E Wilen. 2003. Economic impacts of marine reserves: the importance of spatial behavior. Journal of Environmental Economics and Management, 46: pp. 183 – 206.

[71] Sobel J. 1993. Conserving biological diversity through marine protected areas, a global challenge. Oceanus, 36: pp. 19 – 26.

[72] Stewart R R and H P Possingham. 2005. Efficiency, costs and trade-offs in marine reserve system design. Environmental Modeling and Assessment, 10: pp. 203 – 213.

[73] Sumaila, U R. 2002. Marine protected area performance in a model of the fishery. Natural Resource Modeling 15 (4): pp. 439 – 451.

[74] Sumaila, U R. 1998. Protected marine reserves as fisheries management tools: a bioeconomic analysis. Fisheries Research, 37: pp. 287 – 296.

[75] Sumaila, U. R., A. T. Charles, 2002. Economic models of marine protected areas: an introduction. Natural Resource Modeling 15 (3): pp. 261 – 272.

[76] Sumaila U R, S Guénette, J Alder, et al. 2000. Addressing ecosystem effects of fishing using marine protected areas. ICES Journal of Marine Science, 57

(3): pp. 752 – 760.

[77] Sumaila U R, S Guénette, J Alder, et al. 1999. Marine protected areas and managing fished ecosystems. Chr. Michelsen Institute, Bergen, Norway.

[78] Smith, M. D., 2004. Fishing yield, curvature and spatial behavior: implications for modeling marine reserves. Natural Resource Modeling 17 (3): pp. 273 – 298.

[79] Stewart, R. R., and H. P. Possingham, 2005. Efficiency, costs and trade-offs in marine reserve system design. Environmental Modeling and Assessment 10: pp. 203 – 213.

[80] Tuck G N and H P Possingham. 2000. Marine protected areas for spatially structured exploited stocks. Marine Ecology Progress Series, 192 (1): pp. 89 – 101.

[81] Wallace S S. 1999. Evaluating the effects of three forms of marine reserve on northern abalone populations in British Columbia, Canada. Conservation Biology, 13: pp. 882 – 887.

[82] Walters C J. 2000. Impacts of dispersal, ecological interactions, and fishing effort dynamics on efficacy of marine protected areas: how large should protected areas be? Bulletin of Marine Science, 66 (3): pp. 745 – 757.

[83] Wantiez L, P Thollot, and M Kulbicki. 1997. Effects of marine reserves on coral reef fish communities from five islands in New Caledonia. Coral Reefs, 16: pp. 215 – 224.

[84] Williams I D and N C Polunin. 2000. Differences between protected and unprotected reefs of the western Caribbean in attributes preferred by dive tourists. Environmental Conservation, 27: pp. 382 – 391.

[85] Willis T J, R B Millar, R C Babcock, et al. 2003. Burdens of evidence and the benefits of marine reserves: putting Descartes before des horse? Environmental Conservation, 30 (2): pp. 97 – 103.

[86] Willis T J, D M Parsons, and R C Babcock. 2001. Evidence of long-term site fidelity of snapper (*Pagrus auratus*) within a marine reserve. New Zealand Journal of Marine and Freshwater Research, 35: pp. 581 – 590.

The Literature Review of Marine Protected Areas: An Economic Perspective

Liu Kang

【Abstract】As a precautionary tool of integrated coastal zone management, marine protected areas (MPAs) are widely adopted to deal with the stresses of marine ecosystem, such as marine environmental pollution, loss of biodiversity, depletion of resources, and habitat damage. The objectives of MPAs are both ecological and socio-economic one, of which the socio-economic one is the basis for its success, and one of the major research fields in MPAs. Both the empirical and normative studies have supported the ecological benefits of MPAs, but strong arguments for its economic benefits, including its fishery and recreational values. Based on the meta-analysis of MPAs literatures, we found that the long-term fishery and recreational benefits, and short-term fishery costs of MPAs co-existed. Most of existing economic literatures are based on bioeconomic models of MPAs, including meta-population model, spatial economic model, deterministic model, and stochastic model. Many results of those studies support the economic benefits of MPAs, but with high level of variety and uncertainty.

【Key Words】Marine protected areas Economic benefit Bioeconomic model Cost-benefit analysis

JEL Classification: Q25, Q57

中国海洋产业可持续发展：
基于主流产业经济学视角的分析

于谨凯 李宝星[*]

【摘要】目前，中国海洋经济可持续发展的研究方兴未艾，本文根据主流产业经济学的框架建立全新的海洋产业可持续发展模型，从产业组织、产业关联、产业结构、产业安全以及产业政策几个方面来全面分析中国海洋产业可持续发展的理论和现实问题，以求构建中国海洋产业可持续发展研究的分析框架，并对海洋产业市场结构、投入产出、政策体系、结构优化等问题进行了深入研究。

【关键词】海洋产业 可持续发展 SCP范式 系统论

一、相关理论文献综述

海洋产业可持续发展的研究与海洋、海洋经济、海洋产业的经济与发展理论是紧密相连的，本文从海洋可持续发展理论、海洋经济可持续发展规划、海洋产业经济理论以及最新的生态足迹理论四个方面进行理论文献的回顾与综述。

---

[*] 于谨凯，博士，中国海洋大学经济学院副教授，青岛：200071，Email：yujinkai8@126.com；李宝星，中国海洋大学经济学院硕士研究生。

（一）海洋可持续发展理论

海洋可持续发展理论是一个逐渐形成的过程。1954 年，为实现海洋渔业的持续捕捞，格拉德（H. S. Gordon）提出了开放式资源的经济模型理论。国外学者肯尼思－博尔丁（Kenneth Boulding）于 1966 年提出，陆地资源危机和海洋资源的相对充足使经济增长新空间理论在海洋开发利用中得到利用和发展，促成了海洋可持续发展理论的形成。1992 年里约联合国环境与发展大会通过的《21 世纪议程》中指出的：海洋是全球生命支持系统的基础组成部分，是人类可持续发展的重要财富。至此，海洋可持续发展理论正式形成，海洋可持续发展理论的研究也逐步兴起。国内学者张德贤指出："海洋可持续发展包括三层含义：海洋经济的持续性、海洋生态的持续性和社会的持续性"。王诗成提出："海洋的可持续发展以保证海洋经济发展和资源永续利用为目的，实现海洋经济发展与经济环境相协调，经济、社会、生态效益相统一"。蒋铁民也指出："海洋开发可持续发展包含保证海洋经济增长的持续性，保持良好海洋生态的持续性和良好的社会持续性"。

（二）海洋经济可持续发展规划

自 20 世纪 80 年代以来，美国、日本、英国、法国和德国等国家分别制定了海洋科技发展规划，提出有限发展高新技术的战略决策。1986 年美国率先制定《全球海洋科学规划》，指出海洋是地球上"最后开辟的疆域"，并于 1990 年发表了《90 年代海洋科技发展报告》，明确提出以发展海洋科技来满足对海洋开发不断增长的需求，以便继续保持和增强在海洋科技领域的领导地位。同年，英国海洋科技发展协调委员会发表了《90 年代英国海洋科技发展战略报告》，提出要优先发展对海洋发展有战略意义的高新技术，并于 1995 年制定了《英国海洋科学技术发展战略》。1997 年日本政府制定了面向 21 世纪的《海洋开发推进计划》，提出利用海洋科技加速海洋开发和提高国际竞争力的基本战略。澳大利亚也出台了《澳大利亚的海洋科学技术计划》。

我国海洋经济可持续发展规划与研究，从 20 世纪 80 年代中期就开始跟踪国际相关研究的动向。1994 年中国率先在全球制定了《中国 21 世纪议程》，确立了中国未来的发展要实施可持续发展战略。1996 年颁布的《中国海洋 21 世纪议程》，提出了中国海洋产业可持续发展战略。2003 年 5 月，国务院批准实施《全国海洋经济发展规划纲要》，是我国政府为促进海洋经济综合发展而制定的第一个具有宏观指导性的文件,对于我国加快海洋资源的开发利用,促进沿海地区经济合理布局和产业结构调整,努力促使海洋经济各产业形成国民经济新的增长点有着重要意义。

(三) 海洋产业经济理论

在海洋产业组织的形成与强化方面。威尔斯等(Niels-jseeberg-elverfeldt, 1997)探讨了私人行业在实行波罗的海共同综合环境方案计划的作用;尼尔森和韦德斯曼德(Nielsen and Vedsmand, 1997)探讨了基于丹麦经验的渔业管理中渔民组织的角色,认为在未来世界渔业将会遇到来自管理体制、技术和市场的挑战,需要实施统筹管理,进而分析了渔民组织的作用;斯伯施特罗姆(Sjostrom, 2004)作了关于海洋运输卡特尔的文献综述,就以班轮公会的形式而结成的合谋关系进行了分析,分析了垄断性卡特尔和破坏性竞争两个模式。

全球化与海洋产业集群升级方面。切蒂(Chetty, 2002)以产业集群理论为框架,对新西兰海洋产业集群演化与国际竞争力提升进行了动态关联分析;理祖卡(Lizuka, 2003)探讨了全球标准与智利大马哈鱼产业集群可持续性问题。

海洋产业结构调整方面。常(Kwaka Yoob Chang, 2005)利用投入产出分析方法研究了海洋产业在韩国国民经济中的作用,认为国内外环境的改变和海洋科技发展需要人们对海洋产业的重新认识,进而要求研究者提供可靠的海洋产业信息;道森(Dawson, 2006)对美国实施个体捕鱼配额后,大比目鱼行业的垂直整合问题进行了研究,认为大比目鱼行业的垂直结构已经产生了明显的变化。

(四) 生态足迹理论

生态足迹概念是由加拿大生态经济学家威廉·里斯(William. Rees)和他的学生瓦克内格尔(Wackernagel)在20世纪90年代提出。瓦克内格尔先后对温哥华、英国、荷兰、澳洲等16个国家和城市进行了生态足迹的实证研究,并在1997年提出了有关世界各国生态足迹的报告。世界自然基金会从2000年开始,在《地球生态报告》中开始采用生态足迹来评价全球及各国的资源利用和生态情况。

我国从1999年开始利用生态足迹理论,国内的学者大多是利用生态足迹进行生态系统评估,通过对不同地区生态足迹的计算来说明对自然资源的利用情况，综合资源、环境、经济、社会等方面研究生态足迹和社会发展的问题。近年来，生态足迹理论已开始运用于海洋资源的可持续开发研究中，具有重大的理论指导意义。

二、基于产业视角的海洋经济可持续发展模型构建

根据张德贤先生的海洋可持续发展理论（2001），影响海洋经济可持续发展的因素主要有经济水平、海洋环境，海洋资源以及技术进步。这几个因素之间相互影响、互相制约与海洋经济的可持续发展密切相关，随着技术水平的不断提高，人们对于海洋的开发不断扩大，再加上陆源污染物大量排入海洋，使得海洋的资源与环境问题开始逐渐暴露出来，环境的恶化和资源的逐渐短缺，基于可持续发展的考虑必须发展环保产业，注意保护环境和合理利用资源，而要发展海洋环保产业，离不开财力和物力的支持，这就需要经济的发展，反过来，环保产业的发展又能促进海洋经济和环境的协调发展。与此同时，技术水平的提高，可以提高人们保护环境和合理利用资源的能力。

通过对资源、环境资源的代际配置①以及技术进步的推进作用的分析，可以推得实现海洋经济可持续发展的均衡点的松弛条件，也可以得到关于环境与技术在可持续发展中的均衡条件，即单位财富用于享受所增加的边际效用与用于治理污染使下一代得到的边际效用折现是相同的，以及在考虑技术进步时的单位财富用于享受所增加的边际效用与用于技术进步使下一代得到的边际效用折现相同。在模型中，主要考虑经济、环境质量、资源以及技术这四个要素，其中海洋产业经济水平是由产业组织、产业结构、产业政策几个方面共同决定的。其动态控制模型为：

$$\max \sum_{n=1}^{\infty} U_n [\pi_n(O, S, P) - a_n - b_n - d_n, S_{1n}, S_{2n}, E_n, T_n] \delta^n \quad (1)$$

$$S_{1n+1} - S_{1n} = g_n(S_{1n}, E_n) - R_{1n} \quad (2)$$

$$S_{2n+1} - S_{2n} = h_n(a_n, S_{2n}, T_n) - R_{2n} \quad (3)$$

$$E_{n+1} - E_n = p_n(T_n)F_n - \eta E_n - e_n(b_n, T_n) \quad (4)$$

$$T_{n+1} - T_n = t_n(d_n, T_n) - kT_n \quad (5)$$

$$F_n = F_n(R_{1n}, R_{2n}, E_n, T_n) \quad (6)$$

---

① 资源代际配置即资源在各代人之间的配置，是指每一代人在经济发展的同时，如何确定当代的资源使用量，以及如何给后代留下适宜发展的资源数量。

$$\pi_n = p_n F_n - c_n(R_{1n}, R_{2n}, E_n, T_n) \qquad (7)$$

其中，R_{1n}，R_{2n}，S_{2n}，a_n，b_n，$d_n > 0$，$\pi_n - a_n - b_n - d_n > 0$，$S_{1n} > \bar{S}$；$U_n(\pi_n - a_n - b_n - d_n, S_{1n}, S_{2n}, E_n, T_n)$ 表示第 n 代人的效用函数，效用函数中，用 $\pi_n(O, S, P)$ 表示社会净利润，其中 O、S、P 分别表示产业组织、产业结构、产业政策；$\pi_n - a_n - b_n - d_n$ 表示第 n 代的可支配收入，a_n 表示用于寻找和发现不可再生资源的费用，b_n 表示用于治理污染的费用，d_n 表示用于技术进步的费用；S_{1n} 表示第 n 代时可再生资源的存量，S_{2n} 表示第 n 代时不可再生资源的存量，E_n 表示第 n 代时环境污染水平，它指排放污染物积累总量，T_n 表示第 n 代人具有的知识技术水平。δ 表示社会贴现因子，目标函数为追求人类效用最大化。假设效用函数是可微的，显然，它具有性质：$pU_n > 0$（其中 $n = 1, 2, 3, 4, 5$ 表示对第 n 位变量求偏导）。

公式（2）~（5）分别表示再生资源存量、不可再生资源存量、环境质量、技术进步的约束方程，$g_n(S_{1n}, E_n)$ 表示再生量，R_{1n} 表示第 n 代使用的再生资源量，它作为控制变量；$h_n(a_n, S_{2n}, T_n)$ 表示投资为 a_n、技术水平为 T_n 时的不可再生资源的发现量，R_{2n} 为第 n 代使用的不可再生资源量，a_n 和 R_{2n} 作为控制变量；$e_n(b_n, T_n)$ 表示技术水平为 T_n、投资 b_n 时第 n 代污染的减少量，第 n 代污染总量为 $p_n F_n$（p_n 为单位产品产生的污染量），η 表示环境自净能力；k 为知识折旧（淘汰）系数。等式（7）中 p_n 表示第 n 代产品的单价，F_n 为生产函数，c_n 表示第 n 代生产的成本。

通过建立汉密尔顿（Hamilton）函数，并进行最大化求导后，可以得到均衡点处的松弛条件为：

$$\zeta \geq 0, \quad \zeta(\pi - a - b - d) = 0, \quad \varepsilon \geq 0, \quad \varepsilon(S - \bar{S})$$

这样，海洋产业可持续发展的分析就可以分为产业组织、产业结构、产业政策等几个子系统，根据复杂系统论，我们可以建立海洋产业可持续发展的系统模型，本文从各个子系统之间的协调和整体协调来建立系统模型。

1. 子系统之间的协调发展

海洋产业可持续发展的子系统之间的协调发展可以通过分析各子系统指标之间的相互影响，然后计算子系统之间的相互影响，进而考察子系统之间的协调发展情况。具体模型通过以下步骤来说明：

步骤1：分析各子系统指标之间的相互影响

设 β_{ij}^{pq} 表示第 i 个子系统的第 j 项指标受第 p 个子系统的第 q 项指标的影响系数，其中 $i, p = 1, 2, \cdots, n$；$j = 1, 2, \cdots, mi$；$q = 1, 2, \cdots, np$。

β_{ij}^{pq} 可以通过专家定性分析或关联分析得到，并使 $\beta_{ij}^{pq} \in [-1, 1]$。

$\beta_{ij}^{pq} > 0$ 表示正的影响（促进作用）；$\beta_{ij}^{pq} < 0$ 表示负影响（抑制作用）；β_{ij}^{pq}

=0 表示无影响。

特别地，$\beta_{ij}^{pq}=1$，并且 $\beta_{ij}^{pq}\neq\beta_{pq}^{ij}$，当 $i\neq p$，$q\neq j$ 时。

步骤 2：计算各子系统指标受其他子系统的总影响

用 β_{ij}^p 表示第 p 个子系统所有指标对第 i 个子系统的第 j 项指标的影响：

$$\beta_{ij}^p = \sum_{q=1}^{p} \beta_{ij}^{pq} X_{ij}$$

从而可以得到，其他 $m-1$ 个子系统对第 i 个子系统的第 j 项指标的总影响可以表示为：

$$\beta_{ij} = \sum_{\substack{p=1\\q\neq 1}}^{m} \sum_{q=1}^{n_p} \beta_{ij}^{pq} X_{pq}, \quad i=1,2,\cdots,n; j=1,2,\cdots,m_i$$

步骤 3：子系统指标受其他子系统的综合影响即子系统指标间协调发展系数

子系统指标受其他子系统的综合影响即子系统指标间协调发展系数 D_{ij} 可用下式给出：

$$D_{ij} = \beta_{ij} \Big/ \sum_{p=1}^{n} \sum_{q=1}^{m_p} \beta_{ij}^{pq}, \quad i=1,2,\cdots,n; \quad j=1,2,\cdots,m_i$$

步骤 4：子系统之间的协调发展系数

在求得了 D_{ij} 的基础上，就可以计算子系统之间的协调发展系数 D_i：

$$D_i = \sum_{j=1}^{m_i} D_{ij} X_{ij}, \quad i=1,2,\cdots,n$$

根据海洋经济可持续发展子系统内部的协调发展系数和子系统之间的协调发展系数，就可以考察整个海洋产业可持续发展系统协调发展的状况。

2. 整体协调发展指数

定义海洋产业可持续发展的整体协调发展指数为 D：

$$D = \sum_{i=1}^{m} \alpha_i (\mu_{i_1} U_i + \mu_{i_2} D_i)$$

其中权重满足：$\sum_{i=1}^{m} \alpha_i = 1$，$\mu_{i_1} + \mu_{i_2} = 1$

模型说明：模型本身对于数据虽然没有特殊要求，但是对目标值和观测值或预测值均为大于 0，并应该进行无量纲化处理；α_i、μ_{i_1}、μ_{i_2} 均为系数，一般可以取大于 0 的值。

根据上面建立的模型，本文海洋产业可持续发展是从产业组织、产业结构、产业政策等几个方面来分析的，从主流产业经济学角度来分析海洋产业可持续发展中各个要素的运作机理，如图 1 所示。

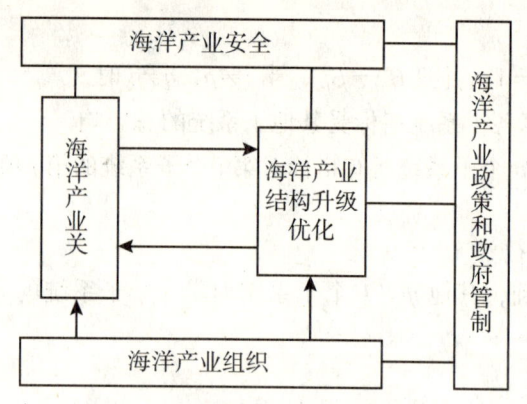

图1 海洋产业可持续发展系统

如图1所示，海洋产业组织状况是海洋产业可持续发展的组织基础，其组织形式和发展状况能够在很大程度上决定海洋产业的发展潜力和发展模式；海洋产业关联互动和海洋产业结构的优化升级，使产业关联与产业结构在海洋产业可持续发展中提供巨大动力的形式，通过关联性互动使整个海洋产业形成链条式发展，为海洋产业的可持续发展提供广阔的发展平台，而海洋产业结构的升级优化，能够大力推进海洋产业的可持续发展向前推进；海洋产业安全与海洋产业政策是海洋产业可持续发展的有力保障，海洋产业安全能够保证在经济全球化趋势下海洋产业可持续发展的国家属性，而海洋产业政策的制定和政府管制的进入能够为海洋产业各个方面的可持续发展提供政策支持和引导，并规范海洋产业可持续发展的总体方向。

三、中国海洋产业可持续发展的组织基础：产业组织分析

海洋产业组织是海洋产业能够实现可持续发展的根本因素和基础，在海洋产业的发展中，其产业组织的形式和特点决定了海洋产业的发展状况和发展潜力，一个好的产业组织形式和内容可以很大程度上实现海洋产业资源配置的优化，人力资本等资源的充分利用等，可以说海洋产业组织形态和状况是决定中国海洋产业可持续发展的基础。

首先，海洋产业市场结构的形态决定了海洋产业的资源配置状况，而资源的优化配置与否是海洋产业能够实现可持续发展的重要标志，资源的优化配置可以使得在海洋产业内部的资源得以充分的调动和使用，从而创造最好的运行效果。一般来说，越是接近完全竞争的市场结构特点，越能够充分优化配置资

源，而完全垄断的资源配置效率最为低下。

其次，根据传统的 SCP 分析范式，产业的市场行为决定与产业的市场结构，这种传统的观点随着产业组织理论研究的深入，已经渐渐发生改变，海洋产业内企业的市场行为同样可以对市场结构发生反作用，它受到政策、制度等方面的影响，其本身也成为海洋产业能够可持续发展的组织基础的重要一个方面。

此外，海洋产业进入壁垒、规模经济等外生条件对于海洋产业的可持续发展同样具有重要的影响作用。

（一）中国海洋产业市场结构分析

海洋产业市场结构是指海洋市场主体的构成，海洋市场主体之间的相互作用及相互联系。在海洋产业市场上，进行商品交换的主体是具有独立经济利益的海洋企业和个人。这些市场主体在市场中的地位、规模和数量比例关系，以及它们的生产技术特点，它们在市场上交换的特点构成了具体的市场。

在不同的市场结构条件下，企业间的竞争内容、竞争特点、竞争强度均有所不同。海洋产业市场结构形态有：完全竞争、垄断竞争、寡头垄断、完全垄断。一般来讲，海洋第一产业的市场形态接近完全竞争，而第二产业中的制药、化工等生产性海洋产业接近垄断竞争形态，在第二产业中的船舶制造和海洋第三产业中的运输业等属于寡头垄断产业，而海洋油气等矿产开采业接近于完全垄断行业。我国各个海洋产业的市场结构分类如表 1 所示。

总之，不管是从我国的实际还是国际的现状来看，海水养殖业和近海捕捞业，也就是通常所指的海洋第一产业存在大量的渔民散户经营，同时小规模的养殖和捕捞企业也非常多，这使得这两个海洋产业的市场结构形态接近于完全竞争；我国的海洋制药业、海洋化工业、滨海旅游业等海洋产业具有明显的垄断竞争特征，在这些海洋产业中，不存在具有支配性力量的海洋企业，市场价格由市场供求的状况来决定。同时这些海洋产业还具有地域分布非常广泛的特点，这和海洋产业的特征是联系在一起的，即具有市场分割性。中国的海洋运输业最具有明显的寡头垄断特性，我国只有中国海洋运输集团公司和中国远洋运输集团公司两家超大规模的海洋运输企业，其他很多小的海运公司是通过租赁这两家公司的船只来运作，即使加上马士基等外国的海洋运输企业，中国海洋运输市场上 90% 以上的份额被不到 5 家的企业占据，可以说是非常明显的寡

表1　　　　　　　　　　　海洋产业市场结构形态分类

市场形态	完全竞争	垄断竞争	寡头垄断	完全垄断
海洋产业	海洋渔业（包括海洋捕捞业和养殖业）	海洋制药业、海洋化工业、滨海旅游业等	海洋运输交通业、海洋船舶业等	海洋油气业、深海采掘业
海洋企业数量及规模	数量很多并且规模很小，即大量的渔民散户经营	数量众多，每个企业的产量在产业总产量中只占一个较小的比例	少数几家大海洋企业占有大部分市场份额	只有唯一一个海洋企业，规模庞大
海洋产品差别	完全相同没有差别，生产同质产品	存在着差别，但是相互之间的替代弹性较大	有的有差别，有的基本没有差别	在产品市场与陆域企业构成寡头垄断，产品差别小
海洋企业进出状况	进出无障碍	进入壁垒较低，新海洋企业能够自由进入产业	进入障碍比较高	相关陆域企业进入壁垒低，新企业进入壁垒极高
信息共享情况	接近完全信息	由于海洋产业特点，构成众多分割市场	不流通信息	无

头垄断市场。此外，海洋造船业、海洋工程建筑业也具有类似的市场形态特征；在我国，海洋油气业中只有中国海洋石油公司一家企业，这种市场状态就是完全垄断市场。在这样的海洋产业市场状态中，作为唯一具有生产和经营能力的企业，其特征与完全垄断厂商的特征并不完全一样，因为在相同产品的市场上，存在相同产品的陆域企业，其并不能成为市场唯一的供给者。

（二）中国海洋产业市场行为及市场绩效分析

1. 中国海洋企业市场行为的经济特点分析

中国海洋企业的市场行为特点从追求规模效益、区域内的集聚效应、易被国家政策引导等几个方面来阐释。

（1）一般企业的市场行为特点。海洋企业首先是一个在市场经济条件下运行的行为主体，具有市场行为主体的一般特性。价格行为是它们进行市场活动和参与市场竞争的主要行为和工具。海洋企业会根据所处市场和竞争情况的不同，采用不同的定价手段，如在垄断市场，为了限制对手企业的进入，会采用限制性定价，把价格定低，从而取得防止对手企业进入的目的，等等。

（2）追求规模效益。海洋产业整体上是一个注重规模经济效应的产业，即使是海洋第一产业的海洋捕捞和养殖业，要想获得更高收益，以及在进行远洋捕捞等行为时，都需要有大的规模企业来运作，海洋的第二、第三产业更是如此。因此对于海洋产业来讲，天生具有追求大规模生产，形成企业内部规模经济效应的特点。这主要是因为海洋是一个人类并不能天生适应的环境，进行海洋开发需要有工具的帮助，海洋开发进行得越深入，需要的技术和装备水平就越高，这无疑大大增加了企业的成本，要想降低这些成本并顺利进行海洋开发，就需要企业在一个大的规模上运行，这样有利于均摊固定成本，从而取得收益最大化。

（3）区域内的集聚效应。海洋企业基本上都聚集在海洋沿岸的区域内，同时由于不同地区的海洋特性和海洋企业间的依赖性，使得不同区域内的海洋企业又形成集聚的趋势和特点。我国的山东半岛是中国渔业发展最为集中的地区，海洋化工业主要集中在华东沿海地区，而我国的"长三角"主要是海洋第三产业发展迅速。这样的海洋企业区域集聚特性，有利于发挥该海洋产业整体的规模经济效应，企业之间可以共同享有信息、开发技术，从而促进海洋企业及其所属产业的发展。

（4）处于易被国家政策引导状态。我国的海洋企业中，海洋基础工业等大型的海洋企业大部分都是国有企业，同时这些国有企业的发展需要国家资金和政策的大力扶持，才能在国际竞争中取得有利地位，因此对于国家政策的依赖性很大。另一方面，其他的海洋企业受这些基础性海洋企业以及国家海洋基础设施的影响很大，所以也要被动地跟随国家政策的引导。整体来说，由于海洋产业的市场化运作机制不健全以及海洋产业的特点，导致海洋企业的行为受到国家政策的影响比较大，处于被引导的状态。

（5）购并、产品差别化行为特征不明显。中国海洋油气业、远洋运输业、深海采掘业等产业天生就具有垄断或寡头垄断的特点，因此并没有进行购并的基础和前提，所以海洋产业通过购并而形成效率垄断的行为并不明显。海洋技术的研发和产业化基本都是在国家控制体制内运行，因此企业在细分市场上的产品差别化行为也不明显。这些非价格市场行为的缺失，也从一定程度上反映了中国海洋产业市场运行体制的不完善。

2. 中国海洋产业绩效评价准则

在海洋产业的绩效评价中，应该采取的准则与陆域产业的评价方法有着密切的关联，都是以提高社会福利水平作为主要依据。这包括海洋产业的资源配置效率，海洋产业的技术效率，海洋企业的规模效率，海洋企业的组织管理效率等。

(1) 海洋产业的资源配置效率。产业资源配置效率是用来评价市场绩效最基本的指标，海洋资源的有效配置能使海洋产业资源得到充分的利用，进而可以提高整个产业的运行效率和福利水平。一般认为，完全竞争的市场机制可以保证最优的资源配置效率。在我国海洋产业中，由于海洋产业对于规模经济和技术水平的高要求，不能一味地追求完全竞争的市场结构，在促进海洋资源有效配置的过程中，应该遵循有效竞争①的原则，这样有利于海洋产业绩效的提高和资源的配置。在海洋产业发展中并不一定要形成完全竞争的市场结构，在竞争过度的行业要促进优胜劣汰，提高企业的规模经济效益；在竞争不足的寡占型市场要促进有效竞争的产生，促进资源的有效配置、技术进步以及企业效率提高。

(2) 海洋产业的技术效率。海洋产业的技术密集特征不仅体现在大多数的海洋产业都是高科技产业，即使是海洋产业的海水养殖和捕捞这样的海洋第一产业，也需要科技的带动推进其发展。所以海洋产业的技术效率对于海洋产业的绩效表现影响非常重大。海洋产业的技术效率不仅指海洋产业的技术水平状况，更包括其技术进步情况。技术进步是产业和企业经营绩效的源泉，它对于海洋产业的市场结构、企业的投资效率、劳动投入效率等都有重要影响。对于海洋产业技术效率的评价，我们可以采用 SFA 模型②和 DEA 方法③进行分析，本文采用的是科技贡献率测算方法（王芳、王文等，2002）。

(3) 产业内企业的规模及状况。由于海洋产业发展高投资、高技术的要求，海洋企业规模经济效应的发挥对于海洋产业的绩效表现有着至关重要的作用。海洋产业内企业达到规模经济的规模水平，摆脱规模不经济的状况，充分发挥企业的生产能力对于中国海洋产业的发展有着决定性的作用。我国海洋产业中很多产业的企业数量很少，只有一家或少数几家国有企业，属于垄断或寡头垄断市场，在国有垄断海洋企业中，由于代理成本、激励成本以及管理成本等方面的成本增加和资源浪费，存在比较严重的 X——非效率状况。

(4) 产业进入壁垒状况。产业的进入壁垒影响到产业的运行绩效，新企业的进入会增加产业的竞争，从而使得价格和利润下降，进而科技使得资源配

---

① 有效竞争是指能够使企业获得合理利润，并实现技术创新良性循环的竞争条件。
② SFA 模型的基本模型包括：$\ln(y_{it}) = a_{it} + \beta_{it}\ln(L_{it}) + \gamma\ln(K_{it}) + v_{it} - u_{it}$；$TE_{it} = exp(-u_{it})$；$u_{it} = \beta(t)u_i$；$\beta(t) = \exp[\eta(t-T)]$；$\gamma = \frac{\delta_{it}^2}{\delta_{it}^2 + \delta_{iT}^2}$。其中 i 表示第 i 个产业，t 表示时间序列，y、L、K 分别表示产出、劳动和资本投入，TE_t 表示技术效率水平。
③ DEA 方法是指数据包络分析，其实质是根据一组关于输入、输出的观察值来估计有效生产的前沿面并与之进行多目标综合效果评价，适用于多输出同类决策单元的有效性评价等。

置更加合理，同时新企业的进入可以提高产业的生产率以及促进技术创新扩散。

产业的进入壁垒可以分为结构性、策略性和体制性三种，我国海洋产业的壁垒主要是结构性和体制性的壁垒。像油气、矿产以及海水淡化、海水利用、海洋电子通信等公共事业，就是由于我国的体制等原因而绝大部分由国有企业经营。而像船舶制造业这样的海洋基础工业近年来虽有大量的民间资本进入，但也只是集中在比较普通的民用船只，而科技含量比较高的军用船只和科研考察船只还是由几大国有造船厂来制造。当然这里存在由于投资量巨大只能由国家来承担的原因，但像这样国有企业垄断的局面并不利于这些产业向更高层次发展，适当和有效的竞争，明晰的产权对于产业和企业的发展非常有利，在考虑市场容量的基础上，我国海洋产业应该多引入竞争机制，从产业体制和政策方面作出改变和调整。

（5）海洋产业投资效率。投资对于国民经济各个产业的发展都有着非常重要的作用，特别是在我国主要依靠投资拉动经济增长的大环境下，投资对于海洋产业的影响也非常大。投资效率主要是指在海洋产业发展中，投资的运行效率、投资方向以及投资数量满足海洋产业可持续发展的程度。

近年来，我国海洋产业发展势头良好，海洋产业投资逐年增加，所以利用好对于海洋产业的投资就显得尤为关键，投资效率可以从以下几个方面进行分析：一是对其他海洋产业的发展有着广泛的影响，能满足不断增长的市场需求并由此而获得较高的和持续发展速度的主导海洋产业的投资状况；二是各个区域的投资协调情况；三是对于高新技术海洋产业的风险投资管理和运营情况；四是投融资渠道的顺畅与否。

四、中国海洋产业关联分析

产业关联的实质是经济活动过程中各产业间的技术经济联系。海洋产业的可持续发展不可能是单一的产业的独立活动，这不可能实现可持续发展。只有通过与关联产业形成链条式的联动发展，才能为海洋产业提供巨大的发展空间，为海洋产业的可持续发展提供足够的动力，并且有利于形成协调、稳定、快速的发展模式。海洋产业关联分为两个方面，一是各个海洋产业之间的前后向关联，二是海洋产业与陆域产业的关联，这两个方面是相辅相成又不完全相同。海洋产业之间的关联是产业部门内部的投入产出关系，前后相关联紧密，对于产业的运行具有重要的支持作用；陆域关联产业一般是海洋产业的基础性产业或高级前向关联产业，能够为海洋产业的升级、优化提供巨大的支持作用。

（一）海洋产业关联方式

海洋产业关联的方式也有多种分类方法，主要有前向关联和后向关联、单向关联和环向关联、直接关联与间接关联的区分。

1. 前向关联和后向关联

海洋产业前向关联是指海洋某一产业为其他海洋产业或非海洋产业提供产品投入的关系，例如对于海水养殖业来说，它与海洋水产品加工业之间的关系便是前向关联关系；海洋产业后向关联是指海洋某一产业在生产过程中需要其他海洋产业或非海洋产业提供中间产品作为该海洋产业要素投入的关系，例如对于海洋运输业来说，它与海洋造船业之间的关系便是后向关联关系。

2. 单向关联和环向关联

海洋单向关联是指海洋产业部门之间或者海洋产业与非海洋产业部门之间按先后顺序排列的中间产品投入的关系。即海洋某一先行产业部门为后续产业部门提供产品，以供其生产消费，但后续产业部门的产品不再返回先行产业部门的生产过程中。而环向关联是指海洋某一产业部门依据前、后向的关联关系组成了产业链。例如海洋捕捞业与海洋水产品加工业就属于单向关联的方式，而海洋能电力工业与滨海电站建筑业就是环向关联的关系，前者为后者提供电力投入，后者又为前者提供发电设备。在现实的经济生活当中，产业与产业间的关系多是环向关联的关系，海洋产业也不例外。

3. 直接关联与间接关联

直接关联是指经济活动过程中两个海洋产业部门之间或者海洋产业与非海洋产业之间存在着直接的提供产品、提供技术的联系。间接关联是指以上两个产业部门本身不发生直接的生产技术联系，而是通过其他产业部门的中介作用才有间接的联系。举例说明，海洋造船业为海洋交通运输业提供船舶，这是直接关联；海洋造船业与水产品加工业表面上看没有直接联系，然而海洋造船业为海洋捕捞业提供船舶，海洋捕捞业又为海洋水产品加工业提供鱼虾，所以海洋造船业与海洋水产品加工业的关系就是间接关联的关系。需要指出的是，在现实的海洋经济活动中，海洋产业之间以及海洋产业与非海洋产业之间的关系多表现为既存在直接关联的关系，又存在间接关联的关系。因此在考虑某一海洋产业对整个海洋产业领域存在的影响时，既要考虑到它与其他产业的直接关联关系，还要充分考虑到它与其他产业所有的间接关联关系。

海洋产业关联的这几种方式不是独立而言的，这些概念可以交叉重合。例如，海洋捕捞业与海洋水产品加工业之间的关系，既是单向关联又是直接关联，并且海洋捕捞业是海洋水产品加工业的前向关联产业。

(二) 海洋产业关联效应

在产业链中的各海洋产业，都不能独立的存在，必须通过供给与需求的关系与其他海洋产业或非海洋产业发生联系。各海洋产业相对于其后向关联产业而言，成为要素的供给者，通过向其他海洋产业或非海洋产业提供要素的投入来确立自己在产业链中的地位；相对于其前向关联产业而言，成为市场的需求方，通过对其他海洋产业或非海洋产业产出的消费来显示其在产业链中的作用。海洋直接产业关联效应是衡量海洋产业间或海洋产业与其他非海洋产业间直接前向或后向关联程度的指标。海洋产业直接前向关联效应的大小用前向关联指数来反映，海洋产业直接后向关联效应的大小用后向关联指数来反映。

具体公式如下：

$$L_{F(i)} = (\sum_{j=1}^{n+m} x_{ij})/x_i, \quad i=1, 2, \cdots, n$$

其中，$L_{F(i)}$表示海洋 i 产业的前向关联指数，x_i 为海洋 i 产业的全部产出，x_{ij} 为海洋 i 产业对 j 产业提供的中间投入。j 产业既包括海洋产业，也包括非海洋产业（从第 1 到第 n 产业属于海洋产业，第 $n+1$ 到第 $n+m$ 产业属于与海洋产业有关联关系的非海洋产业）。

$$L_{B(j)} = (\sum_{i=1}^{n+m} x_{ij})/x_j, \quad j=1, 2, \cdots, n$$

其中，$L_{B(j)}$表示海洋 j 产业的后向关联指数，x_j 为海洋 j 产业的全部产出，x_{ij} 为海洋 j 产业消耗 i 产业的中间产品。i 产业既包括海洋产业，也包括非海洋产业（从第 1 到第 n 产业属于海洋产业，第 $n+1$ 到第 $n+m$ 产业属于与海洋产业有关联关系的非海洋产业）。

利用这种方法也可以计算与海洋产业有关联的非海洋产业的前向关联效应和后向关联效应，算法与海洋产业相同，只需要将第一个公式中的 i 和第二个公式中的 j 从 $n+1$ 取到 $n+m$ 即可。即

$$L_{F(i)} = (\sum_{j=1}^{n+m} x_{ij})/x_i, \quad i=n+1, n+2, \cdots, n+m$$

为非海洋产业 i 的前向关联效应；

$$L_{B(j)} = (\sum_{i=1}^{n+m} x_{ij})/x_j, \quad j=n+1, n+2, \cdots, n+m$$

为非海洋产业 j 的后向关联效应。

同理，$L_{F(i)} = (\sum_{j=1}^{n} x_{ij})/x_i, \quad i = n+1, n+2, \cdots, n+m$

为非海洋产业 i 对于海洋产业的前向关联效应；

$L_{B(j)} = (\sum_{i=1}^{n} x_{ij})/x_j, \quad j = n+1, n+2, \cdots, n+m$

为非海洋产业 j 对于海洋产业的后向关联效应。①

（三）海洋产业关联的投入产出分析

海洋产业间或者海洋产业与其他非海洋产业间存在着错综复杂的关联关系，这种关联关系正是通过海洋投入产出表体现出来的。海洋投入产出表采取棋盘式，纵横交叉，能从生产消耗和分配使用两个方面来反映海洋产业间或海洋产业与陆域产业间的运动过程，反映海洋社会产品的再生产过程。

海洋产业不是一个孤立的产业，从大的系统上来看，海洋产业可以作为国家整个国民经济的一个重要部门，它对国民经济起着举足轻重的作用，因此海洋诸产业与非海洋产业之间存在着相互消耗和相互提供产品的内在联系，即投入产出关系。

同时，海洋产业又可分为各个不同的产业部门，例如海洋捕捞业、海水养殖业、海洋水产品加工业、海洋交通运输业、海洋盐业、滨海旅游业、海洋石油与天然气业、滨海砂矿业等，又可按照三次产业划分法分为海洋第一产业、海洋第二产业、海洋第三产业，海洋产业内部各产业部门间也存在着密切的投入产出关系。

基于海洋产业的这些特点，我们在编制海洋产业投入产出表时，一定要注意以下两点：第一，产业的划分要全面。不仅要包括海洋第一产业、第二产业和第三产业，还要包括与海洋产业密切相关的非海洋产业，即陆域产业。第二，海洋产业和非海洋产业的区分要清晰明了。最好把海洋产业和非海洋产业区分开，先排列海洋产业，再排列非海洋产业，这样在计算反映海洋产业之间或海洋产业与陆域产业之间的各种系数时，才会有更清晰的认识。

表2是一张简化的一般形式的海洋投入产出表，主要部分有六个［表中（一）~（六）］：（一）反映海洋产业的生产仍用于海洋产业生产所消耗的产品，即海洋产业内部的投入产出；（二）反映海洋产业生产用于非海洋产业即陆域产业生产所消耗的产品；（三）非海洋产业提供给海洋产业消耗的产品；

---

① 海洋产业关联的前向关联效应类似于海洋产业结构分析中的中间需求率，后向关联效应类似于中间投入率。

（四）与海洋产业相关的非海洋产业部门之间的投入产出；（五）反映海洋产业和非海洋产业用于积累、消费和出口的最终产品；（六）反映海洋产业和非海洋产业在生产过程中所消耗的固定资产、劳动报酬和社会纯收入等，可合计为毛附加值。

表2　　　　　　　　　　一般形式的海洋投入产出

投入＼产出		海洋产业中间需求 1, 2, …, n	非海洋产业中间需求 1, 2, …, k	最终需求 积累，消费……，出口
海洋产业的中间投入	1, 2, …, n	（一）	（二）	（五）
非海洋产业的中间投入	1, 2, …, m	（三）	（四）	
毛附加值		（六）		
合计				

本文限于篇幅将不对海洋产业关联的投入产出进行细化和实证分析，这里只是提出投入产出分析的一般工具和方法。

五、海洋产业结构及其升级优化：基于海洋产业可持续发展的产业结构优化模型

本文通过建立基于海洋产业可持续发挥的结构优化模型来说明产业结构升级优化对于海洋产业可持续发展的影响和作用。

1. 目标函数

我国海洋产业的可持续经济增长与产业结构优化模型从经济、社会与环境三个方面设计了目标函数。①

（1）经济增长目标。海洋产业结构优化的目的之一就是促进海洋经济的增长与发展，而增长作为经济发展最主要的目标也就成了本优化模型的一个主要的目标函数，模型确定报告期的 GDP 与基期的 GDP 之比达到最大，即

$$\max \theta_1 = \frac{i^T[X(t_m) - A(t_m)X(t_m)]}{i^T[X(t_0) - A(t_0)X(t_0)]}$$

---

① 潘文卿：《一个基于可持续发展的产业结构优化模型》，载《系统工程理论与实践》2002年第7期。

式中，X 为各海洋产业总产出列向量；A 为投入产出直接消耗系数矩阵；t_m、t_0 分别表示报告期与基期；i^T 为以 1 为元素的列向量的转置，即为求和算子。

（2）充分就业目标。海洋经济的增长是为了促进社会的发展，劳动力就业是社会存在的主要问题，为此本优化模型的目标函数中包括了使失业率尽可能小的充分就业目标：

$$\min \theta_2 = \left[1 - \frac{i^T X(t_m)/l(t_m)}{L(t_m)}\right]$$

式中，$L(t_m)$ 表示 t_m 年劳动力得总供给量；$l(t_m)$ 为 t_m 年以海洋产业总产值核算的全员劳动生产率。

（3）污染控制目标。经济、社会、环境的协调发展是可持续发展的主要内涵，保护生态环境、合理开发与利用海洋自然资源将是中国进入 21 世纪后长期所面临的主要问题之一，我们所建立的优化模型考虑了环境保护的污染排放最小化问题。

$$\min \theta_3 = \sqrt[m]{\frac{(U^K)^T X(t_m) F(t_m)}{(U^K)^T X(t_0) F(t_0)}} - 1, \quad K = 1, 2, \cdots$$

式中，$U^K = (U_{i1}^k, U_{i2}^k, \cdots, U_{in}^k)$ 为各海洋产业第 i 种能源消耗所产生的第 K 种污染物的排放系数，$i = 1, 2, 3$ 分别表示煤炭、燃油和天然气，$K = 1, 2, \cdots$，表示各种污染物；$F(t_m) = f_{i1}(t_m), f_{i2}(t_m), \cdots, f_{in}(t_m)$ 为各海洋产业第 i 种能源的消耗系数。

2. 约束条件

经济运行主要受各经济变量间的内在联接关系决定，即现实的经济运行机制决定了模型的结构，主要的经济运行机制约束为：

（1）投入产出平衡约束。

$$X(t_m) - A_m X(t_m) \geq Y_c(t_m) + Y_I(t_m) + Y_{EM}(t_m)$$

式中，$Y_c(t_m)$ 表示最终消费列向量，它由居民消费列向量 $Y_{c1}(t_m)$ 与政府消费列向量 $Y_{c2}(t_m)$ 组成；Y_I 表示资本形成列向量；Y_{EM} 表示净出口列向量。

（2）生产能力约束。

$$X(t_m) \leq \beta(t_m) K(t_m)$$

式中，$\beta(t_m)$ 是以各海洋产业资本存量产出率为元素构成的列向量；$K(t_m)$ 是由各海洋产业资本存量为元素的列向量。

（3）消费需求约束。

$$[1 - s(t_m)]\{i^T[X(t_m) - A(t_m)X(t_m)]\} \geq i^T Y_c(t_m)$$

式中，$s(t_m)$ 表示 t_m 年的国内储蓄率。

(4) 资本形成约束。
$$[s(t_m) + s_f(t_m)]\{i^T[X(t_m) - A(t_m)X(t_m)]\} \geq i^T Y_I(t_m)$$

式中，$s_f(t_m)$ 表示国外资本流入（外国储蓄）占 GDP 的比重。

(5) 净出口约束。
$$s_f(t_m)\{i^T[X(t_m) - A(t_m)X(t_m)]\} \geq i^T(-Y_{EM})$$

(6) 海洋自然资源约束。

海洋自然资源的约束主要考虑到了其再生性与适度开发与利用的相互适应问题。
$$X_i(t_m) \leq V_i(t_m)$$

式中，$V_i(t_m)$ 表示海洋产业部门产出所能达到的最高限。

(7) 非负约束。
$$X_i(t_m) \geq 0$$

3. 模型设定

该模型是一个多目标规划问题。首先将 GDP 增长目标变换为最小化问题：
$$\min \theta_1 = \frac{i^T[X(t_0) - A(t_0)X(t_0)]}{i^T[X(t_m) - A(t_m)X(t_m)]}$$

然后，我们分别设定三个目标的权数 $\lambda_i(i=1,2,3)$，得到线性加权目标函数：
$$\min(\lambda_1\theta_1 + \lambda_2\theta_2 + \lambda_3\theta_3)$$

六、我国海洋产业安全分析

在经济全球化趋势不断发展的背景下，国家之间的经济联系越来越紧密，这种紧密的关系不仅带来了经济利益，同时也容易造成经济的不安全，特别是对于发展中国家，更容易被发达国家控制经济命脉。本文对海洋产业安全的研究是以产业安全的基本理论、观点、评价体系作为基础，着重分析海洋产业安全的影响因素等内容。最后，对我国海洋产业的形势和趋势进行了分析。

我国 2004 年发表的《全国海洋经济发展规划纲要》开篇指出，"中国是海洋大国，管辖海域广阔，海洋资源可开发利用的潜力很大。加快发展海洋产业，促进海洋经济发展，对形成国民经济新的增长点，实现全面建设小康社会目标具有重要意义。"因此，进行对中国海洋产业安全的客观评价，找出中国海洋产业安全的弱势所在并且提出相应的解决方法，对我国经济增长至关重要。

（一）海洋产业生存和发展环境指标

根据前文建立的海洋产业安全评价体系，海洋产业生存和发展环境是否安全要从海洋产业的融资环境、海洋产业生产要素环境、海洋产业市场需求环境和海洋产业发展环境四个方面综合评价。

1. 海洋产业的融资环境评价

（1）资本效率指标。首先，大量的海洋高技术企业由于规模小、成熟时间短且正处于成长的初期，风险投资者因为缺乏退出渠道，心怀疑虑，使这些企业获得银行信贷的难度加大，丧失了融资机会。其次，风险资本要想从产权交易市场获得资本回收及增值，由其他企业分担风险也极其困难。目前在产权交易市场进行交易的成本远远高于股票市场的成本，过高的税费标准使退出成本提高而增加了投资风险。同时，目前全国还不存在一个统一的产权交易市场，实行跨行业、跨地区的产权交易困难，更降低了成功的可能性。[①] 再次，我国投向海洋产业的资金分散在财政、科技、经济、计划等部门和银行，利用效率低，无法形成资金使用合力。综合看来，我国海洋产业融资困难，资本效率较低。

（2）资本成本指标和资产负债率指标。作为融资主渠道的传统的银行信贷融资方式不能适应海洋产业尤其是高科技资金的投入运作规律，导致海洋产业主体企业通过银行贷款来获取资本的成本太高，使得原本有竞争力的企业背上沉重的负担，存在产业竞争力安全隐患。一般说来，海洋产业的高科技项目拥有巨大市场前景，但由于产品未形成产业化，无法预估投资回报的前景，而且许多项目承担企业，特别是民营高科技企业自身没有大的固定资产，短期之内又无法通过内部积累获取资本，所以导致这些企业的资产负债率过高，其对资金的需求使得以银行为中介的融资体制成本过高，使产业的发展水平和竞争力水平受到融资环境的制约。

2. 海洋产业生产要素环境评价

（1）劳动力成本指标。虽然我国劳动力要素丰裕，劳动力成本相对较低，但因物耗过大，管理费用偏高，一定程度上抵消了劳动成本优势。以中水集团斐济金枪鱼钓船项目为例，在其总成本中，人工费用占17.6%，管理费用占7%，其他占75.4%；在扣除外方销售代理费和运输费用后的总收入中，人工费占16.8%，管理费占6.5%，利润占4.3%，如不扣除代理和运费，利润仅

[①] 高忻：《海洋产业如何流过融资瓶颈》，载《中国投资》2004年第6期。

占销售收入的2.3%。金枪鱼钓船净资产收益率为3%~7%。就我国台湾省金枪鱼项目来说，因其大量雇用大陆船员，在其总成本中，人工费用占27%，管理费占1.7%，其他占71.3%。金枪鱼钓船一般净资产收益率达8%~11%。而在日本金枪鱼船队中，人工成本却高达销售收入的30%~40%。①

（2）劳动力素质指标。目前我国参与海洋产业的劳动力素质普遍偏低，平均熟练程度低，文化程度不高。以远洋渔业为例，大部分船员仅为小学或初中毕业生，高级职务船员人才更是缺少。而日本、韩国等国家外派职员船员多为中专以上，并受过专门的职业培训，无论在捕捞技术、经营管理和外语水平方面，均比我们要高出一头。虽然我国劳务费用相对便宜，但劳动生产率较低。

（3）知识资源指标。从技术方面来看，我国海洋技术与国外先进海洋国家相比，海洋油气勘探和开发技术、海洋造船技术等海洋技术差距在5年左右，海水直接利用技术差距10~15年，海洋捕捞技术、海洋能利用技术差距则在15年以上。据有关专家统计分析，目前发达国家科学进步因素在海洋经济发展中的贡献率已达到80%左右，而我国则在30%左右，科技进步在我国渔业经济增长中所占的比重目前约为46%。②

以海洋勘探技术为例，海洋勘探资料是主张国家海洋权益、进行海洋划界的基础，法律赋予的权利只有在掌握充分勘探资料的基础上才能变成实实在在的利益，否则只能沦为空谈。然而，我国目前对专属经济区和大陆架的勘测范围不到一半，目前的勘查工作还不能完全满足其管辖海域海洋资源开发和申请外大陆架的要求。目前关于"区域"资源的争夺更为激烈，海洋勘探对维护中国在国际海底区域的海洋权益也至关重要。海洋勘探资料是申请矿区的重要依据，是掌握"区域"话语权的利器。

3. 海洋产业市场需求环境指标

据联合国有关组织估计，世界人口的60%居住在距海岸100公里的沿海地区。结合中国的人口分布和流动规律分析，2020年我国沿海地区人口总数将达到6亿~7亿。沿海地区将形成包括基础产业、高新技术产业、旅游业等多种海洋产业在内的多元化的经济模式，这就在食物、土地、水资源、矿产、交通、能源等多个方面带来巨大的需求压力，其中包括滩涂土地利用、海运货运需求、港口装卸、海运船舶，以及海水浴场、海盐、淡水、石油和其他战略性矿产也都有巨大的需求。海洋产业需求前景广阔，在缓解经济社会发展的需

---

① 中国水产（集团）总公司、中国水产科学研究院WTO课题组：《WTO对我国远洋渔业的影响和对策研究》，载《中国渔业经济》2002年第1期。
② 广东省海洋渔业"十一五"规划研究课题组：《全球海洋经济及渔业产业发展综述》，载《新经济》，2005年第8期。

求方面将作出越来越大的贡献。①

4. 海洋产业发展环境指标②

（1）海洋产业总产值。1979年中国海洋产业总产值为64亿元，1989年上升至245亿元，10年间上升了近4倍，1999年中国海洋产业总产值3 651.30亿元，比1989年上升了14.9倍，2003年中国海洋产业总产值首次突破万亿元大关，达到10 077.71亿元，如图2所示。

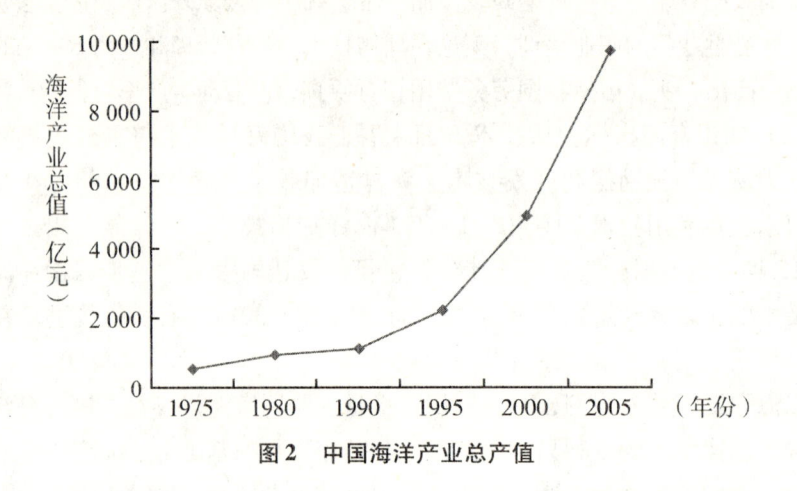

图2 中国海洋产业总产值

（2）海洋产业增加值。1996年中国海洋产业增加值为1 266.30亿元，占整个沿海地区国内生产总值的3.3%，占全国国内生产总值的1.9%，到2003年中国海洋产业增加值为4 455.54亿元，相当于全国国内生产总值的3.8%，高于同期国民经济的增长速度，如图3所示。

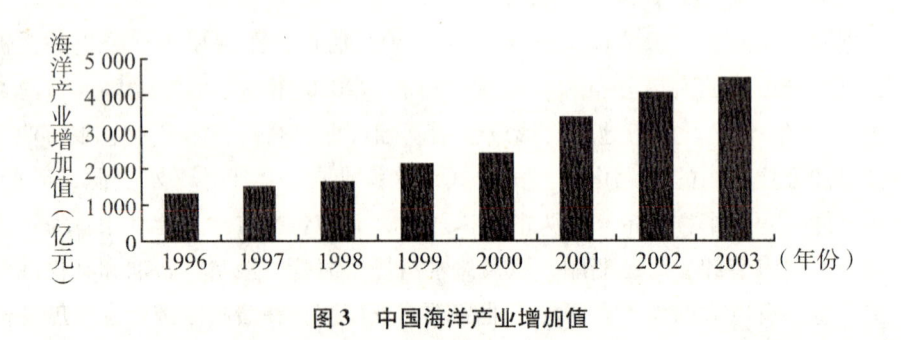

图3 中国海洋产业增加值

---

① 杨金森：《2020年的中国海洋开发》，http：//www.soa.gov.cn/zhanlue/hh/9.htm。
② 王永生：《我国海洋产业评价指标及其测算分析》，载《海洋开发与管理》2004年第4期。

（二）中国海洋产业控制力安全评价

从我国 2004 年修订的外商投资产业指导目录可以看出，中国政府对海洋产业控制力安全给予了极大重视。

（1）在鼓励外商投资产业目录中仅限于：渔业中名特优水产品养殖、深水网箱养殖；海洋油气业中的石油、天然气的风险勘探、开发，但限于合作；低渗透油气藏（田）的开发，限于合作；提高原油采收率的新技术开发与应用，限于合作；物探、钻井、测井、井下作业等石油勘探开发新技术的开发与应用，限于合作；海运中定期、不定期国际海上运输业务（规定外资比例不超过 49%）和国际集装箱多式联运业务（规定外资比例不超过 50%，不迟于 2002 年 12 月 11 日允许外方控股，不迟于 2005 年 12 月 11 日允许外方独资）；高科技新兴产业中海洋开发及海洋能开发技术、海水淡化及利用技术、海洋监测技术。

（2）限制外商投资产业目录中有海运业的水上运输公司，外资比例不超过 49%。

（3）禁止外商投资产业目录中包括：渔业中的我国稀有的珍贵优良品种的养殖、种植（包括种植业、畜牧业、水产业的优良基因），我国管辖海域及内陆水域水产品捕捞。

因此，在一段时期内，中国在海洋产业实行外资管制，对海洋产业拥有绝对的控制力和决策权，海洋产业在控制力方面是安全的。

（三）中国海洋产业竞争力安全评价

中国海洋产业发展境况各不相同，产业竞争力也存在差别，由于在前文中关于产业竞争力问题有过详细描述，这里只简要介绍一下我国渔业、船舶制造业和海洋油气业的产业竞争力安全。

（1）渔业竞争力安全。我国水产品产量区域集中度高。据国家统计局的统计数据，水产品全国总产量 84.11% 集中在 10 个省份，在这 10 个省份中产量排名前五位的省份的海水产品占各省总产量比重均在 50% 以上，利润比重 26.14%，这与韩国前五名船厂约 80%~90% 的集中度相差甚远。其中山东、福建、浙江、辽宁海水产品比重高达 80% 以上，浙江高达 93.16%，产量增长幅度较高的为辽宁、湖南、江苏、广西，增长幅度均在 6% 以上，辽宁高达 8.7%。

从生产价格指数观察，2005年全国渔业产品生产价格指数105.14%，比畜牧业产品高4.62个百分点，比种植业高3.59点，比农产品生产价格总指数高3.75点，渔业产品的比较效益非常明显。①

(2) 船舶制造业竞争力安全。中国船舶制造业的市场集中度是属于中集中度的寡占型市场，并且很大程度上是国家、政府产业政策扶持的，这是因为政策与法律扶持制造的进入壁垒限制了新企业的进入，才提高了市场的集中度，而并不是市场规律运作的结果。虽然我国船舶产业的行业集中这几年正在不断提高，但与发达国家相比还有差距，韩国前五家造船企业的绝对集中度大于90%，日本的产品也主要收缩集中于三大造船厂，②我国船舶制造业竞争力安全有待进一步提高。

(3) 海上油气业竞争力安全。在封闭条件下，中国石油产业一直保持一种极高寡占型的市场结构。重组前，中国石油天然气总公司和中国海洋石油总公司垄断上游的石油天然气开采业，产量占全国的比重为99.16%；中国石油化工总公司垄断下游石油业，加工量占全国的比重为81.24%；三大公司实行上下游分割垄断，彼此之间的竞争十分微弱。重组后，三大集团公司实现了上下游一体化经营，业务上相互交叉，打破了上下游分割垄断的局面。但由于石油、石化、海洋分别位于北方、南方及海上，形成了地域分治的格局。同时，三大集团公司的油气生产和油气加工量仍占全国的98.62%和94.17%，垄断依然存在。所以，在不考虑进出口的前提下，从总体产业层面上分析，石油产业是一种极高寡占型的市场结构。③

(四) 中国海洋产业权益安全评价

从中国国家管辖海域范围及公海权益维护安全来看，我国海洋权益安全正受到严峻的挑战。

按照《联合国海洋法公约》的规定，中国拥有近三百万平方公里的可管辖海域，在世界海洋大国中名列第九位。此外，中国作为国际海底资源开发的先驱投资者之一，在太平洋公海海域还拥有75万平方公里的海底矿区专属开发权。

但是，在我国濒临的渤海、黄海、东海和南海这些海域内，不少区域都与相邻、相向周边国家存在划界矛盾，同时与一些国家还存在着岛屿的主权争端。

---

① 吴湘生：《中国渔业产业区域竞争力分析与未来发展战略》，载《北京水产》2006年第5期。
② 王连军：《中国船舶制造业：SCP范式分析》，载《重庆工商大学学报》2005年第6期。
③ 白雪峰、王宇奇：《开放条件下国石油产业垄断初探》，载《科技与管理》2003年第6期。

1. 存在海域划界矛盾与主权争端

（1）在东海，中国与日本、韩国存在划界问题。20世纪70年代，日、韩两国在东海划定大面积的大陆架，称为"共同开发区"。但是，根据《联合国海洋法公约》对大陆架的有关条款的规定，东海大陆架是中国的自然延伸，面积77万平方公里的海区中应归我管辖的为54万平方公里，是中国的领土，因此东海大陆架在所谓的共同开发区中存在着中、日、韩三国的主张重叠区。按日本的无理要求，日本与中国有16万平方公里、韩国与中国有18万平方公里的争议地区。中国对钓鱼岛及其附属岛屿拥有主权，但由于钓鱼岛周围海域资源丰富，日本一直处心积虑霸占为己有。

（2）南海海域不仅拥有丰富的油气资源，南海渔业资源也具有极高的经济价值。此外，南海地区在矿藏、旅游、运输等方面也都具有极高的价值。但是有关沿海各国不顾中国提出的"搁置争议、共同开发"的主张，疯狂侵占我国的岛礁，大肆掠夺我国资源，严重破坏了我国海洋产业权益安全。我国南沙、中沙和西沙群岛的诸多岛屿，更是被菲律宾、马来西亚、越南、印度尼西亚、文莱等国家瓜分。南沙群岛由数百个岛、礁、沙洲、沙礁组成，其中适合驻军的条件较好的岛礁除太平岛外，均被越、菲、马占领。仅被越南侵占的就有29个，被菲律宾占领的有9个，被马来西亚控制的有5个。

海洋资源被掠夺则更为严重。到20世纪90年代末期，周边国家已经在南沙海域钻井一千多口，发现含油气构造二百多个和油气田180个，1999年年产石油4 043万吨、天然气310亿立方米，分别是中国1999年全年近海石油年产量和天然气产量的2.5倍和7倍。自1981～2002年，越南已从南沙海域的油田中开采了1亿吨石油、15亿多立方米的天然气，获利250亿美元。南海石油已成为越南国民经济的第一大支柱产业。马来西亚、菲律宾和文莱也大肆开采石油，文莱通过开采南沙石油，由穷国变成了富国。目前，南沙周边国家已在南沙海域与数十家外国公司联合打井超过一千口，年产石油数千万吨，天然气数百万立方米。此外，南沙海域海底还蕴藏着丰富的矿产，如铜、镍、钴、锰以及其他稀有金属。[①]

（3）在黄海总面积38万平方公里的海域中应划归中国管辖的有25万平方公里，但在海域划界问题上韩国主张等距线为界，如果按此划分，韩国则可多划18万平方公里。中国与朝鲜和韩国存在着18万平方公里的争议海区。

（4）由于起步较晚以及深海技术比较落后，中国在公海的远洋开发处于劣势。

[①] 吕建华：《论我国海洋权益的现状与保护》，载《海洋法苑》2003年第6期。

2. 实行海洋产业国际安全合作

从中国实行的海洋产业国际合作来看权益安全，我国的对外合作涉及海洋养殖业、海上运输业、海洋油气业等诸多方面，并且已与二十多个国家签有各种类型和级别的海洋合作协议。这些对外合作都为促进中国海洋科技的进步、培养人才、引进资金，加快海洋经济和海洋产业的发展，起到了积极的促进作用。

（1）中国与俄罗斯、印度签署了海洋合作协议，其范围涵盖了海洋政策法规、海洋生态环境保护、海岸带综合管理、基础海洋学研究、新技术开发、减灾防灾、大洋资源勘探开发和极地考察和研究等合作领域。作为一个发达的海洋大国，俄罗斯在实施全球海洋战略、海洋研究与开发以及资源管理方面水平较高，特别是在大洋资源勘探和开发、极地考察和研究等领域基础雄厚，处于世界领先地位。印度在海洋研究、海洋资源开发和海洋观测技术等方面也有独到之处。我国积极开展与俄罗斯、印度等海洋大国的合作，可以借鉴他们在海洋科学技术和管理方面的经验，有利于促进我国海洋工作的发展。

（2）中国与美国海洋领域合作协议在早期着重于参与海上渔业联合执法，共同查处公海流网作业。之后签订了《中美海洋与渔业科技合作议定书》，并一直召开中美海洋与渔业科技合作联合工作组会议，进一步加强双方在海洋政策、海洋环境与气候、海岸带管理、海洋资料交换和极地方面的合作。

（3）与周边国家的合作更是有益于维护中国海洋权益，解决已有的争议，形成安全的海洋权益形势。2002年，中国与东盟共同签署《南海各方行为宣言》，为维护南海地区和平与稳定、促进在南海开展务实合作奠定了重要的政治基础。《东亚海可持续发展战略》于2003年12月在马来西亚通过，为在东亚地区建立海洋事务合作框架，加强各国在海洋事务方面的合作，促进本地区海洋可持续发展，由中国主导地区海洋事务方面跨出了重要一步，并奠定了重要的政治基础。2004年9月，我国召开的"南海潜在冲突研讨会"，通过了国家海洋局提出的南海数据库项目建议书，对南海合作起到积极作用。2004年10月，在中、日、韩渔业高层会议上，三方决定在技术及研究领域开展合作，并在《联合宣言》指导下落实三方渔业协定，开展渔业资源保护合作。2004年11月，我国与菲律宾签署了在双方争议地共同勘探油气资源的协议。2004年12月与印度尼西亚、马来西亚和菲律宾经过友好磋商，达成在海洋政策、海洋与海岸带资源管理、海洋资源开发、海洋环境保护与保全、海洋观测与防灾减灾、海洋与海岸研究及其相关领域的培训教育与信息交换等方面开展合作的意向；与马来西亚达成在海洋科研调查、海岛开发等领域内开展合作的意向；这对加强与南海周边各国的海洋合作、稳定周边将起到重要作用。

七、中国海洋产业可持续发展的产业政策与政府管制

为了实现海洋经济的可持续发展,就需要政府制定和实施合理有效的海洋产业政策。因此,明确政府在海洋产业政策制定与实施过程中的职能与作用至关重要。

(一) 中国海洋产业政策分析

如今,海洋产业的发展已经引起了中国各级领导、各级政府的高度重视。中央和地方政府也相继出台了相关的海洋发展政策。中国政府于1996年制定的《中国海洋21世纪议程》和2003年制定的《全国海洋经济发展规划纲要》,对于海洋产业政策的制定起到了指导性的作用。一般而言,产业政策主要由产业结构政策、严业布局政策、产业技术政策等方面构成。在本文中则根据产业政策的功能将海洋政策体系划分为支持性、引导性和发展性政策三个模块,如图4所示。

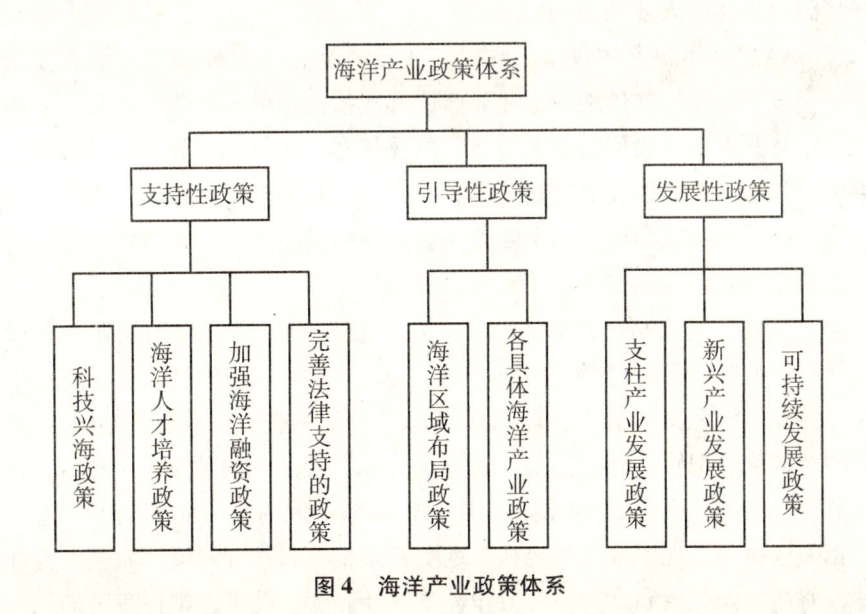

图4 海洋产业政策体系

1. 支持性海洋产业政策体系
海洋产业的发展与基础设施的建设以及相关的配套环境息息相关。支持性

海洋产业政策体系旨在为海洋产业提供一个发展的基础平台，为海洋经济的发展缔造一个良好的外部环境。支持性海洋产业政策主要包括以下主要内容：

（1）科技兴海政策。坚持科技兴海，加强科技进步对海洋经济发展的带动作用，是提高海洋产业竞争力的重要途径。依靠科技进步提高我国海洋经济的竞争力，总的思想是从我国海洋产业发展的实际出发，以引进和吸收海洋科学和技术新成果、海洋科学和技术的联合攻关以及技术成果转化为手段，以海洋开发企业为主体，以高等院校、科研机构为依托，大力发展海洋高新技术，建立和完善海洋科技多层次的研究、开发体系，提高海洋科学与技术水平。各级人民政府对海洋科技能力建设的投入，要重点支持对海洋经济有重大带动作用的海洋生物资源综合开发技术、海水资源开发利用技术、海岸与海洋工程技术、海洋能源及矿产开发应用的新技术、滨海旅游资源的开发技术、海域资源和环境评估技术、海洋监测及海洋灾害预报预警技术、海洋污染防治和生态保护技术等关键技术领域，重点开展科技攻关和成果应用，力争有突破性进展。

（2）海洋人才培养政策。培养海洋科学研究、海洋开发与管理、海洋产业发展所需要的各类人才，是提高科技对海洋经济发展的关键要素。我国必须采取各种手段，健全高等院校的海洋专业，提高对海洋专业的教学、科研资金的投入，加强职业教育，培养多层次的海洋科技与管理人才，满足海洋各个产业和部门对于海洋人才的需求。

（3）增加海洋融资的政策。在投资主体方面，不仅要确立企业在发展海洋经济过程中的投资主体地位，还要鼓励和支持海内外各类投资者依法平等参与海洋经济开发。从投资对象来看，首先要拓宽海洋基础设施建设的投资、融资渠道，为海洋产业的发展奠定基础；其次，一些海洋新兴产业具有较高的投资风险，因此我国还要建立和不断地完善海洋高科技产业风险投资基金，为此类产业的发展提供资金支持。

（4）完善法律支持的政策。目前，我国已经制定和实施的法律和行政法规包括：《中华人民共和国海洋环境保护法》、《中华人民共和国渔业法》、《中华人民共和国海上交通安全法》、《涉外海洋科学调查研究管理规定》、《矿产资源勘查区块登记管理办法》等，这些法律和行政法规的涉及范围较广，内容与《联合国海洋法公约》的原则和有关规定是一致的。在此基础上，进一步完善相关法律、法规体系，制定和组织实施海域权属管理制度、海域有偿使用制度、海洋功能区划制度，完善海洋经济统计制度，既可以维护我国的国家主权和海洋权益，也促进了海洋资源的合理开发和海洋环境的有效保护，使中国的海洋综合管理初步走上法制化轨道，为国内外企业进入海洋经济领域创造良好的投资环境。另外，我国还要建立适应海洋经济发展要求的行政协调机制，明确中央和地方、各有关部门在海洋管理中的工作职责,加强海上执法队伍的建设以及相关法律、法规的执法力度。

2. 引导性海洋产业政策体系

目前,我国各个地区和各个海洋产业之间的经济发展仍存在一定的盲目性,引导性海洋产业政策体系旨在指导海洋经济健康有序的发展,其主要包括:

(1) 海洋区域布局政策。根据我国2003年制定的《全国海洋经济发展规划纲要》,中国的海洋经济区域分为海岸带及邻近海域、海岛及邻近海域、大陆架及专属经济区和国际海底区域。开发建设的时序和布局是:由近及远,先易后难,优先开发海岸带及邻近海域,加强海岛保护与建设,有重点地开发大陆架和专属经济区,加大国际海底区域的勘探开发力度。

(2) 具体海洋产业政策。海洋产业要调整结构,优化布局,扩大规模,注重效益,提高科技含量,实现持续快速发展。加快形成海洋渔业、海洋交通运输业、海洋油气业、滨海旅游业、海洋船舶工业和海洋生物医药等支柱产业,带动其他海洋产业的发展。

3. 发展性海洋产业政策体系

(1) 支柱产业发展政策。坚持突出重点,大力发展支柱产业。努力扩大并提高海洋渔业、海洋交通运输业、海洋石油天然气业、滨海旅游业、沿海修造船业等支柱产业的规模、质量和效益。发挥比较优势,集中力量,力争在海洋生物资源开发、海洋油气及其他矿产资源勘探等领域有重大突破,为相关产业发展提供资源储备和保障。

(2) 新兴产业发展政策。所谓海洋新兴产业,是以海洋高新技术发展和海洋资源大规模开发为背景的,由产业演化形成期进入成长期的海洋产业,它既是指按照海洋产业形成规模开发的海洋产业群体,又是指依据海洋资源开发在相同或相关价值链上活动的各类企业所构成的企业集合。我国要发挥市场配置资源的基础性作用,大力调整和改造传统海洋产业,积极培育新兴海洋产业,加快发展对海洋经济有带动作用的高技术产业,深化海洋资源综合开发利用。

(3) 可持续发展政策。加大海洋环境保护投入,保障海洋经济可持续发展。坚持经济发展与资源、环境保护并举,保障海洋经济的可持续发展。重点加强污染源治理,加快建设沿海城市、江河沿岸城市污水和固体废弃物处理设施。完善海洋生态环境监测系统与评价体系。加强赤潮研究、监控和预报,建立赤潮监控区。鼓励非政府组织开展海洋生态环境保护活动。加强海洋环境保护的国际合作。加强海洋生态环境保护与建设,海洋经济发展规模和速度要与资源和环境承载能力相适应,走产业现代化与生态环境相协调的可持续发展之路。

（二）中国海洋产业可持续发展中的政府职能定位

（1）政府权力的强制性保证了海洋产业政策的顺利执行，使其能够有效地调动、配置海洋资源。政府是因社会公共需要而产生的、对全体社会成员具有普遍性的组织，具有宪法授予的公共权利。依靠这种公共权利，政府可以法律的形式制定产业政策和市场规则，并以司法强制力保证其执行。强制力使政府在引导海洋经济发展的过程中具有独特的力量，可以通过各种法规、政策来规范企业和个人行为，引导他们合理地利用和保护海洋资源。如政府的征税权、禁止权、处罚权和奖励权等。政府借助于宪法赋予的这些权利，可以合法地禁止或允许某些企业或个人采取某种行为或退出某些海洋经济活动；政府通过其掌握的权力优势和信息等资源优势，能够有效地配置与海洋产业相关的各种资本；政府所拥有的庞大的财政实力和独特的财政货币权利，使其在特定条件下，可以通过财政货币权利来加大对从事支柱性与新兴海洋产业的企业的投入，引导企业或个人投身这些海洋产业。所以，政府的权力优势，是其制定与实施海洋产业政策的制度保障。

（2）政府权力的公共性决定了海洋产业政策的制定是以可持续发展为基础的。政府是公共权利的代表，受公众的委托来管理社会，因此，政府应该承担起保护海洋环境和资源、促进海洋产业可持续发展的重任，对海洋经济的可持续发展作出长远规划、统筹解决。政府通过制定一定的产业政策，一方面鼓励企业、个人积极参与海洋产业的生产，另一方面，采取各种措施合理配置资源，防止由于利润最大化的市场原则造成对海洋环境和资源的损害。为此，这既需要政府直接投资进行有利于海洋经济发展的基础设施的建设和基础产业的研究开发，也需要在政府的诱导和协助下或在政府的直接规制下，采用市场手段兴办海洋产业，以满足海洋经济发展的需要。

（3）政府通过制定海洋产业政策对海洋经济进行管理，要能够做到行之有度，注重干预和引导的平等性、科学性和有效性，真正做到"掌舵"而不是"划桨"。传统经济理论发现了市场失灵，从而导致了政府干预的必要性。以布坎南为代表的公共选择理论发现政府干预经济活动也会失灵。他认为政府介入公共产品的分配会出现"寻租"行为，从而影响了政府干预经济行为的效果。尽管政府在作用中也有可能出现"政府失效"的现象，但只要政府的行为遵循市场经济规律，政府依法行政，调控有度，就有可能避免失败。在社会主义市场经济条件下，企业转换经营机制成为市场的主体，政府转变职能成为市场经济宏观调控的主体。因此，政府对海洋开发利用者的活动不应再是单方面的命令、要求，而应以引导者、调节者的身份出现，通过制定与事实合理有效的产业政策，采用税收等间接调节手段，通过建议、劝告、倡导、奖励等方式，为企业提供良好服务和有效监督，引导企业从事海洋产业的生产。

参考文献

［1］ A. Charns, W. W. Cooper, Q. Wei, and Z. M. Huang, Cone Ratio Data Envelopment Analysis and Multi-objective programming, International Journal of System Science Vol. 20, NO. 7, 1989, pp. 1099 – 1118.

［2］ European Journai of Operational Research, Vol. 2, NO. 6, 1978, pp. 429 – 444.

［3］ Advances in Complex System, http: //www. santefe. edu/bonabeau/.

［4］ Alice Huhhard what are sustainable communities? http: //www. sustainable. doe. Gov/index3. htm.

［5］ Alison Gilbert, criteria for sustainability in the development of indicators for sustainable development, Elsevier Science Ltd. , 1996.

［6］张德贤等．海洋经济可持续发展模型及应用研究［J］．青岛：中国海洋大学学报，2001（1）．

［7］郭万达．现代产业经济辞典［M］．北京：中信出版社，1991．

［8］丰志培，刘志迎．产业关联理论的历史演变及评述［J］．温州大学学报，2005（2）．

［9］王海英，栾维新．海陆相关分析及其对优化海洋产业结构的启示［J］．海洋经济，2002（6）．

［10］许长新，陈浩．海洋产业的关联性研究［J］，海洋经济，2002（5）．

［11］周洪军等．我国海洋产业结构分析及产业优化对策［J］．海洋通报，2005（2）．

［12］董伟．澳大利亚海洋产业计量方法［J］．国外海洋开发与管理，2006（2）．

［13］郭越．近年我国海洋产业增加值率统计分析［J］．2004．16．

［14］刘洋等．海洋产业经济的定量分析技术研究［J］．海洋开发与管理，2005（6）．

［15］王芳等．科技进步评价理论与方法在海洋产业的应用［J］．国土资源科技管理，2002（6）．

［16］臧旭恒等．产业经济学［M］．北京：经济科学出版社，2005．

A Sustainable Development Research on China's Marine Industry: Based on the Viewpoint of the Main Industrial Economics

Yu Jinkai, Li Baoxing

[Abstract] Now the sustainable research of our country's marine economy is just emerging. The paper based on the main industrial economics establishes a new model of marine industry sustainable development, and analyses the theory and real problem of China's marine industry sustainable development from some aspects, such as industry organization, industry connection, industry structure, industry security and industry policies, and on the basis the paper wants to establish an analysis frame of China's marine industry sustainable development. The paper makes a deep research on marine industry's market structure, input-output, policy system, and so on.

[Key Words] Marine industry Sustainable development SCP Mode System theory

JEL Classification: Q56, L16

胶州湾围垦行为的博弈分析及保护对策研究

孙 丽 刘洪滨[*]

【摘要】近十几年来，胶州湾及周边进行了大面积的围填海活动，给胶州湾海岸带及其周围海域造成了地貌形态的改变，生态环境的破坏，以及许多动植物的灭绝等危害。本文采用博弈方法对胶州湾各相关利益主体之间的行为关系进行了分析，叙述了这些行为所造成的沿海地区环境的改变和生态的破坏，回顾了已经采取的措施，提出了控制胶州湾围填海、保护生态环境的一些建议。

【关键词】胶州湾 博弈 围填海 保护

一、引言

博弈论（game theory）是研究决策主体的行为发生直接相互作用时的决策以及这种决策的均衡问题的，也就是说，当一个主体，比如说一个人或一个企业的选择受到其他人、其他企业选择的影响，而且反过来影响到其他人、其他企业选择时的决策问题和均衡问题。在下面的分析中，我们将使用到纳什均衡

---

[*] 孙丽，中国海洋大学海洋管理专业硕士研究生，电子邮箱：sunbenane@yahoo.com.cn；刘洪滨，中国海洋大学海洋发展研究院教授，青岛，266071，电子邮箱：hliu@qingdaonews.com。

理论，所谓的纳什均衡指的是这样一种战略组合，这种战略组合由所有参与人的最优战略组成，也就是说，给定别人战略的情况下，没有任何单个参与人有积极性选择其他战略，从而没有任何人有积极性打破这种均衡。[1]纳什均衡是一种完全信息静态博弈，即信息对于博弈双方来说是完全公开的情况下，双方在博弈中所决定的决策是同时的或者不同时，但在对方作决策前不为对方所知的。博弈论在经济学中已经得到了广泛的应用，但在海洋学中运用博弈论方法对海洋行为主体进行研究的案例目前还很少。本文运用博弈论的有关方法对胶州湾围垦行为以及利益主体之间的关系进行分析，可以有效分析当事人和政府的行为原因。

二、胶州湾的重要地位及开发利用现状

胶州湾地处青岛市区内，现存面积388平方公里，对青岛市的社会、经济发展，乃至生存有着及其重要的作用。我们之所以一次次地提到胶州湾的重要性和作用，是因为胶州湾是青岛的母亲湾，有了胶州湾，才有了青岛这座城市，才有了青岛港，形成了环胶州湾产业集聚带，拉动了青岛经济、社会、文化等各项事业的发展。目前青岛正按照"拥湾发展"的思路建设特色鲜明的现代化国际城市，正在实施的大港口、大炼油、大造船、大旅游、大工业、大物流，而所有这些项目都是围绕着胶州湾进行的，可以说胶州湾是青岛的"财富之湾"和"希望之湾"。

历史上的胶州湾地区，由于人口少、生产力低下，对自然资源开发利用的能力与水平有限，客观上对胶州湾的环境和资源影响不大。而当代由于缺乏统筹规划和对权属的有效管理，以致产生开发利用不当和开发过度现象。

从总体来看，胶州湾岸段与水域空间已充分开发甚至过度开发。大规模的围海造地无疑给青岛地方经济的发展带来了商机，但胶州湾水域面积的急剧减少、环境污染加重也向人们发出了警告。

近十几年来胶州湾沿岸的填海造地工程不断上马，使胶州湾水域面积急剧减小，导致潮流、纳潮量等减小，从而使物理自净能力削弱。与此同时，迅速发展和壮大的工业、海运业、养殖业及急剧增加的城市人口，不断将大量污染物质排入湾内。不协调地发展导致胶州湾水体污染程度加剧，严重影响了渔业资源的再生能力和海水养殖业的发展，胶州湾著名的菲律宾蛤仔亦从20世纪80年代的年自然产十万吨左右，发展到现在完全靠养殖维持局面。

在青岛市"十五"时期发展规划中，确定了青岛的城市性质为"东部沿

海经济中心和港口城市",城市的主体功能为"以港口为主的国际综合交通枢纽,国家海洋科研和海洋产业开发中心",城市总体布局为"以胶州湾东岸为主城,西岸为辅城,环胶州湾沿岸为发展组团";确定了"以港兴市"发展战略,以前湾港区、大港港区、黄岛油港区为主体,以沧口水道北港区、小港区及其他地方港区和专业港口为辅助的多功能中心大港,综合年吞吐能力由现在的1亿吨发展到2010年的1.5亿吨。胶州湾空间资源将面临新一轮的开发利用。

工业要有较大发展,城市人口必然会明显增加,胶州湾的环境和资源负荷将会进一步加大。必须合理利用胶州湾宝贵的岸线资源,保护胶州湾生态环境,维护胶州湾的可持续发展。

三、胶州湾岸线及水域面积演变

(一) 岸线变化

历史上,胶州湾海岸形态演变主要包括地质构造变化、海平面升降、水动力(河流动力和海洋动力)及沉积物运移和分配等,它反映着大自然固有的演变规律,非人类意愿所能为,这种演变状态一直延续到20世纪初期。[2] 随着社会的发展,尤其中华人民共和国成立以后,生产力得到了空前发展,人们出自不同的目的和要求,在胶州湾沿岸进行大量的围海造陆,如建设码头、盖厂房、修护岸、造陆连岛、掘虾池和围盐田、倒垃圾等所谓的当代演变。使胶州湾在当代演变过程中水域面积急剧减少,人类活动变成最主要原因。

20世纪初期,胶州湾畔人烟稀少,人类作用对海湾的影响微乎其微。自然因素在70年代以前对胶州湾冲淤演变起着主导作用。河流输沙在1966年以前占绝对优势,是胶州湾淤积和水域面积、体积减小的主要原因;围湾造陆和河流输沙是60~80年代中期胶州湾水域面积减小和淤积的主要原因;而1985年以后,围湾造陆造成了胶州湾水域面积减小和岸线增长,而上游修库筑坝引起的河流来沙量骤减(减小了2个数量级),使自然的作用更加减弱。

20世纪70年代以后,人为因素成为胶州湾岸形变化的主导,各种海岸工程的填海造陆成为胶州湾水域面积减小的主要原因。据相关资料,1971~1988年胶州湾人为减少的水域面积约60.1平方公里(见图1),[3] 其中东部岸段主要由填海造陆、修堤筑港等海岸工程建设造成海湾总水域面积缩小,而东北、西北和西南部岸段主要由围海造田（盐田和养虾池）引起的。

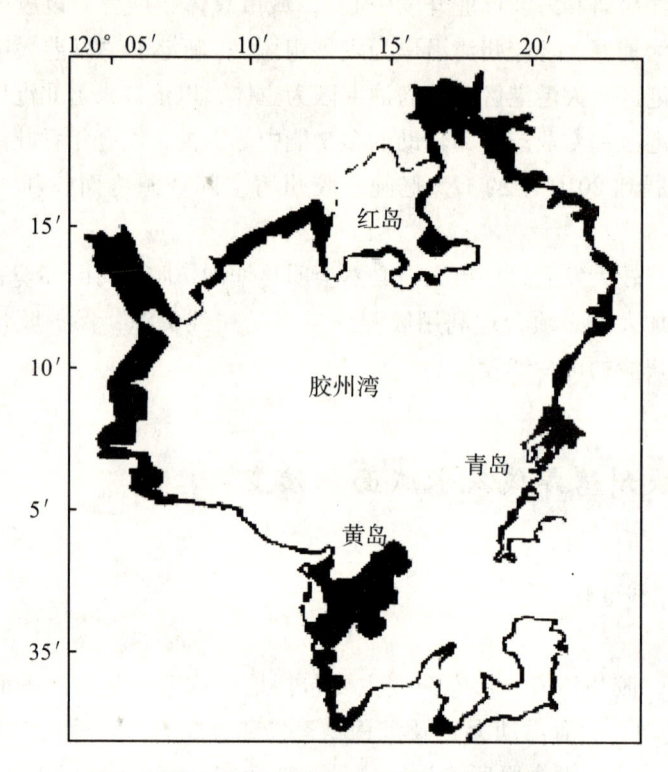

图1　1971～1988年胶州湾岸线与面积动态变化

通过数据收集，1992～2010年人为填海工程造成胶州湾面积减小的主要数据如表1，[4] 胶州湾的岸形变化见图2。

表1　　　　　胶州湾1992～2010年人为占用水域面积　　　　　平方公里

时间段（年）	占用水域名称	占用水域面积
1992～2004	环海公路	7.10
	前湾港	2.5
	黄岛二期油码头	0.5
	安庙码头	0.6
	鲁能仲盛置业公司	1.6
	四方区港	0.8
	航务二公司	0.18
	青岛港集装箱公司	0.3
	远洋公司	1.1

续表

时间段（年）	占用水域名称	占用水域面积
1992～2004	红岛渔港	0.03
	前港保税区	2.1
	海西湾北海船厂	0.33
	四方区临海工业园	1.82
	青岛电厂	2.68
	青岛港黄岛油管六期	0.51
	其他	3.0
	合计	25.15
2004～2010	海洋石油工程	0.68
	海西湾造船基地	1.03
	前湾2010年规划用地	6.41
	其他（估计）	2.0
	合计	10.12
	总　　计	35.27

资料来源：青岛市海洋功能区划办公室，《青岛市海洋功能区划》，2005年。

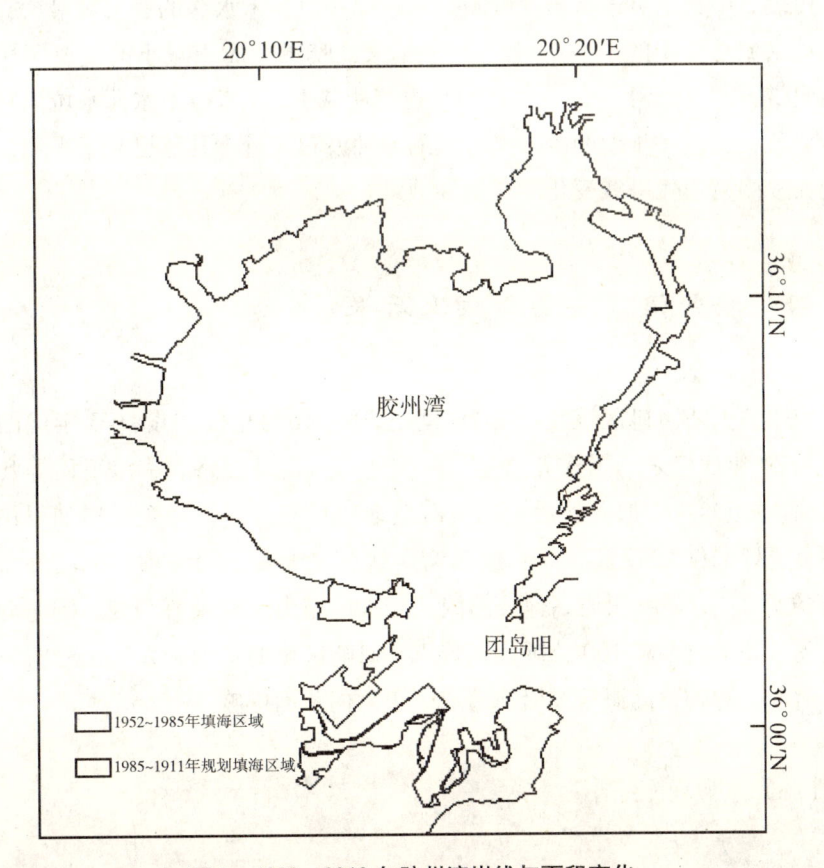

图2　1992～2010年胶州湾岸线与面积变化

资料来源：青岛市海洋功能区划办公室，《青岛市海洋功能区划》，2005年。

(二) 海域不同水深区域变化

按各不同等深线所包围的水域面积进行分析，列出表2。

表2　　　　　　　　　　　不同水深的水域面积分布

年度	范围	潮间带	0~5米	5~10米	10~20米	>20米
1992	面积（平方公里）	85.21	190.17	57.11	29.17	26.46
	百分比（%）	22	49	14.7	7.5	6.8
2002（推算）	面积（平方公里）	20.8	188.63	57.11	29.17	26.5
	百分比（%）	7.9	49.6	15	7.7	6.9

资料来源：青岛市海洋功能区划办公室，《青岛海洋功能区划》，2005年。

由表2中数字可以看出，胶州湾水域面积以5米水深的浅水域最大，它占胶州湾水域总面积的70%以上。近十年来，胶州湾面积减小的主要区域集中在潮间带，0~5米水深的浅水区面积也有所减少，5米以上水深水域面积几乎没有变化，这进一步说明胶州湾自然的冲蚀淤积在近十几年已趋于平衡，人为演变在胶州湾当代岸线变化中是主导因素。

四、胶州湾围垦的影响及后果

胶州湾大沽河口区域经历了20世纪50年代的盐田建设，70年代的填湾造地和80年代以来的围建养殖池、开发港口、建设公路和临港工程等几次填海高潮。[5] 近年来，胶州湾沿岸的围海造地项目和海洋工程数量增加很快，据不完全统计，仅2002年以来，胶州湾内就有二十多个用海项目，填海面积约20平方公里。[6] 环胶州湾高速公路附近较大的填海工程就有两处，围海面积分别达到了267公顷和100公顷。随着人类影响区域的扩大和人工工程向海岸的不断逼近，对胶州湾海域及沿海造成了很大的负面影响。

(一) 水域面积缩小

围垦项目是造成胶州湾水域面积缩小的主要原因。图3是分别对1928年、1958年、1971年、1978年、1985年、1988年、2001年、2003年的水域面积进行的统计，不难看出，从 1958~2003 年的 45 年间，胶州湾面积减少了 173 平方公里，缩小了近 1/3。这是由于 20 世纪 50 年代以后，工、农、渔、盐等行业纷纷向胶州湾"进军"，对胶州湾的开发和利用程度加大，围海造地和各种海洋工程开发的后果，致使胶州湾的生态系统受到严重破坏。

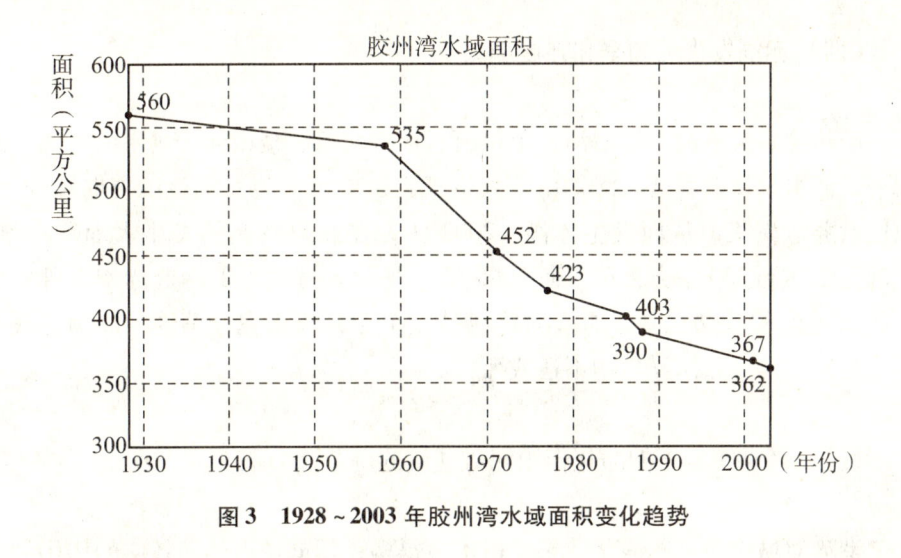

图3 1928~2003 年胶州湾水域面积变化趋势

（二）纳潮量减少

胶州湾水域面积的缩小，其直接后果必然使海湾的纳潮量减少，造成流场改变、水动力强度减弱、水体交换和携沙能力下降。20 世纪 40 年代以前，由河口等注入的泥沙及少量的工业三废等物质，通过涨、落潮，水体交换，几乎全部携带到湾外，胶州湾基本处于稳定平衡状态。1935 年的胶州湾纳潮量为 11.822 亿立方米，1963 年为 10.133 亿立方米，1985 年为 9.144 亿立方米，现在的纳潮量只有 7 亿多立方米。这就是说，纳潮量减少了 4 亿多立方米，只有原纳潮量的近 40%。专家指出，如果听任胶州湾海域面积继续缩小，不远的将来，青岛港将因为泥沙淤积而成为死港，青岛的经济优势将随之不复存在。

（三）海洋生物迅速减少

由于纳潮量减少，污染物得不到疏散，海洋自净能力降低，使胶州湾的污染日益严重，破坏了海洋生态平衡。据 2003 年青岛市海洋环境质量公报显示，胶州湾内中度污染和轻度污染海域约占胶州湾总面积的3/5。受污染影响，胶州湾的生态环境严重破坏，生物多样性迅速减少。20世纪60年代胶州湾河口附近潮间带生物种类多达154种，70年代减到33种，80年代只剩下17种，原有的14种优势物种仅剩下一种，东岸的贝类已不复存在。

（四）赤潮发生的频率和强度加大

胶州湾有大沽河、海泊河、李村河、板桥坊河、楼山河、墨水河等淡水河注入，河水携带营养盐入海，由于人造工程影响了海流及潮流的方向和动力，致使营养盐聚集，从而发生赤潮。2003年，在胶州湾北部发生大面积赤潮，赤潮范围达到200平方公里。而2005年2月和8月发生的两起赤潮事件，污染面积均有80平方公里，且2月赤潮持续1个月仍未完全消退。赤潮造成水体大量缺氧，导致海洋生物不适或死亡。

（五）降低了城市的净化作用和对气候的调节能力

港湾对城市粉尘的净化效果比树木大得多，降低城市热岛效应的作用尤为明显，有关测算表明，城市绿化面积达30%，可以调节温度1℃~2℃，而同样面积的水体可以调节的温度达3℃~4℃。

海洋水可以把春、夏两季储存的热量用于补偿秋、冬两季的亏损，它在季节调节过程中所起的主导作用，使海水温度变化比较平缓。但由于胶州湾的水域面积减少，降低了它对青岛市的气候的调节能力。据专家统计分析，青岛市的气候在变暖，且变暖速度加快，平均最低气温升高幅度大，日较差在缩小，[7]这都与胶州湾的净化作用和气候调节能力减小有关。

五、相关博弈分析

胶州湾围海活动造成的环境恶化等后果已经显现，且这种行为屡禁不止，这都是因为在围海造地的过程中存在着几种复杂关系的博弈，包括开发商与开发商之间的博弈，开发商与政府之间的博弈，以及中央政府与地方政府之间的博弈，等等。

（一）开发商之间的博弈

如果所有的开发商都从海洋可持续发展的角度出发，来保护海洋资源及岸线，这对各方都有好处。但这种情形对一定区域中只有一个开发商来说是可行的，如果这个区域的开发商增加后，要解决彼此共同的问题就不再那么容易了。下面我们将对开发商争抢围垦的行为进行分析。[8]

假定胶州湾共有 n 个开发商，第 i 个开发商围垦面积为 g_i，$i=[1, n]$，$g_i \in [0, \infty)$；$G = \sum_{i=1}^{n} g_i$ 代表 n 个开发商共围垦的总面积；v 代表单位面积的平均价值。我们假设 v 是 G 的函数，$v = v(G)$。由于胶州湾面积有限，所以存在一个最大围垦面积 $G_{\max}$：当 $G < G_{\max}$ 时，$v(G) > 0$；当 $G > G_{\max}$ 时，$v(G) < 0$。在人类进行大力开发之初，当可围垦面积很大时，增加围垦面积也许不会对其他开发商的围垦价值形成太大不利影响，但随着近年来围垦行为的加剧，围垦面积不断增加，围垦的单位面积的价值会急剧下降。因此，

$$\frac{\partial v}{\partial G} < 0, \quad \frac{\partial^2 v}{\partial G^2} < 0$$

在此博弈里，开发商的问题是选择 g_i 以使自己的利益最大化。假定围垦单位面积的土地需要投资 c，则利润函数为：

$$\pi_i(g_1, \cdots, g_i, \cdots, g_n) = g_i v\left(\sum g_j\right) - g_i c, \quad i = 1, 2, \cdots, n$$

最优化一阶条件是：

$$\frac{\partial \pi_i}{\partial g_i} = v(G) + g_i v'(G) - c = 0, \quad i = 1, 2, \cdots, n \tag{1}$$

对此一阶条件可解释为：每增加单位围垦面积有两方面的效应，一方面增加了此围垦面积的价值，另一方面是增加的围垦面积使所有之前的围垦面积的价值下降。

对公式（1）进一步求偏导，

$$\frac{\partial^2 \pi_i}{\partial g_i^2} = v'(G) + v'(G) + g_i v''(G) < 0$$

$$\frac{\partial^2 \pi_i}{\partial g_j \partial g_i} = v'(G) + g_i v''(G) < 0$$

因此，$\dfrac{\partial g_i}{\partial g_j} = -\dfrac{\dfrac{\partial^2 \pi_i}{\partial g_j \partial g_i}}{\dfrac{\partial^2 \pi_i}{\partial g_i^2}} < 0$

就是说，第 i 个开发商的最优围垦面积随其他开发商的围垦面积的增加而递减。仔细观察可以发现，每个开发商在决定增加围垦面积的时候只考虑了对自己利润的影响，而不是对所有开发商利润的影响。

将公式（1）中的 n 个分量进行相加，我们可以得到：

$$v(G^*) + \frac{G^*}{n}v'(G^*) = c \qquad (2)$$

其中，G^* 是纳什均衡的总围垦面积。

社会最优的目的是最大化社会总剩余价值，即 $\max\limits_{G} Gv(G) - Gc$

最优化一阶条件为：$v(G^{**}) + G^{**}v'(G^{**}) = c \qquad (3)$

其中，G^{**} 是社会最优围垦面积。

由公式（2）和公式（3），可以看出 $G^* > G^{**}$，因此，我们得到的纳什均衡是总的围垦面积是大于社会最优的围垦面积的，如果没有政府的强制控制，毫无疑问胶州湾将被无限度地围垦开发。

由此分析，我们可以这样去理解开发商的心理：一方面，当开发商数量增加之后，个人行为的后果对整体来说可能微不足道，但对他自己来说，却影响重大。也就是说，当有人加大围垦开发的力度，而自己却基于永续利用的观点自我克制，那么对整个胶州湾的环境和生态保护作用甚微，但自己却同样要承担环境治理的任务，这时他会选择违规或擦边违规加大围垦；另一方面，如果自己增加围垦面积，那么对整体造成的伤害并不显著，但却可以得到显而易见的好处，这时他选择的策略是增加围垦面积，即在总体上有加大利用空间资源的可能时，自己加大利用而别人不加大利用则自己得利，自己加大利用时其他人也加大利用，则自己不至于吃亏，最终是所有人都加大利用资源直至不能再加大时（再加大利用将全盘崩溃）的纳什均衡水平。

为了不至于达到最终的纳什均衡水平（胶州湾完全被围垦，已再无可开发之地），政府必须作出相应的决策，控制开发商的自主开发行为，将纳什均衡保持在一个社会最优的水平，以达到社会最优开发程度。

（二）开发商与政府间的博弈

至于开发商与政府之间的博弈，我们暂定为开发商与执法监察部门之间的博弈。

假设开发商每增加单位围垦面积增加的收益为 R，监察部门的检查成本为 C，单位面积罚款为 $F(F>C)$，环境成本为 H。[9] 则双方收益如图 4 所示：

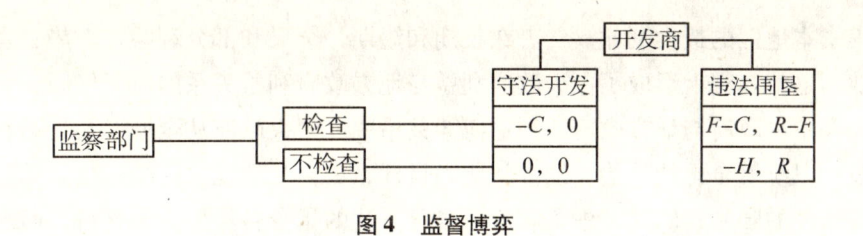

```
                                    ┌──开发商──┐
                           ┌──守法开发  违法围垦
                    ┌─检查─┤   -C, 0    F-C, R-F
          监察部门──┤
                    └─不检查   0, 0      -H, R
```

图 4　监督博弈

设定监察部门检查的概率是 θ，开发商违法围垦的概率是 γ。

给定 γ，监察部门选择检查和不检查的期望收益分别为：

$\pi_G(1,\gamma) = -C(1-\gamma) + (F-C)\gamma$

$\pi_G(0,\gamma) = -H\gamma$

解 $\pi_G(1,\gamma) = \pi_G(0,\gamma)$，得 $\gamma^* = C/(F+H)$。如果开发商违法围垦的概率大于 γ^*，则监察部门应选择检查，反之，不检查。

给定 θ，开发商选择守法和违法的期望收益分别为：

$\pi_p(\theta,1) = 0\theta + 0(1-\theta)$

$\pi_p(\theta,0) = (R-F)\theta + R(1-\theta)$

解 $\pi_p(\theta,1) = \pi_p(\theta,0)$，得 $\theta^* = R/F$。如果监察部门检查的概率大于 θ^*，则开发商很有可能选择不违法围垦，反之，冒险违法围垦。

此混合战略纳什均衡与罚款 F、围垦收益 R、检查成本 C 以及环境成本 H 有关。罚款金额和环境成本越大，开发商违法围垦的概率就越小；检查成本越大，围垦收益越多，开发商违法围垦的概率就越大。因此，一旦开发商选择违法加大围垦力度，如果执法部门同时也加大监察力度，使 θ 足够大，加大惩罚力度 F 足够大，则开发商违法围垦的期望收益 $\pi_p(\theta,0)$ 将小于 0 甚至远小于 0。开发商的目的就是追求最大利润，如果增加其风险成本，那么开发商将无心违法加大围垦。

（三）地方政府与中央政府间的博弈

地方政府与中央政府之间的博弈主要是基于利益关系的博弈。在计划经济时代，地方政府在权力以及利益方面居于明显的从属地位，只能绝对地执行中央政府下达的政策、指示，没有可供其运用的权利资源，因此二者构不成博弈关系。随着市场机制的建立，中央将部分权利下放给了地方政府，在分担中央政府调控压力的同时承担和分享相应的经济责任和利益。中央政府代表国家的整体利益和社会的普遍利益。而地方政府代表局部利益，其执行政策的出发点是谋求本地区的最大利益。二者在长期和短期、全局和部分利益方面势必发生争执，这就客观上形成了中央政府利益与地方政府利益关系的博弈格局。

在中央政府与地方政府的利益博弈关系中，中央政府从全局以及长远利益出发，以统筹全国的政治、经济问题为目标，更会受环境、资源、社会承受能力等条件的限制，因此，中央政府对自身下达的指令自然要全力支持；而地方政府若是完全服从中央意愿，在中央及全局利益得以实现的同时，自身利益可能要受损，特别是在当前注重官员"政绩工程"的情况下，远不如从本地区利益和眼前利益出发来衡量和采取措施，自己安享眼前收益，把可能的问题留给中央。

比如在对待胶州湾的问题上，中央政府从保护环境、治理污染的角度出发，提出对围海造地进行严格管理，《中华人民共和国海域使用管理法》第四条明确规定，"国家严格管理填海、围海等改变海域自然属性的用海活动"，并对围海造地的管理设置了最严格的审批权限，即"填海50公顷以上的项目用海"应当报国务院审批。2002年7月，国务院办公厅专门发出《关于沿海省、自治区、直辖市审批项目用海有关问题的通知》，进一步规定"填海（围海造地）50公顷以下（不含50公顷）的项目用海，由省、自治区、直辖市人民政府审批，其审批权不得下放"。而地方政府顾虑到本地税收问题以及处于保护本地经济的目的，对中央的指令也不尽遵守，致使很多填海工程都是先斩后奏的"三边"工程，即边干、边审批、边论证。这时候地方政府会采取不作为的措施，等待中央政府去对已造成的环境问题采取行动，自己坐享短时间的地区利益，或者报批时采取化整为零的手法躲避中央的监管。

六、研究、规划及管理

在意识到胶州湾的重要性以及对胶州湾的围垦现状有所调查的情况下，政府部门及学者也感到了情况的严重性，并就保护胶州湾采取了一系列措施。

1984年，青岛市科学技术协会组织召开了《胶州湾开发利用综合研讨会》，并出版了论文集，这对胶州湾的开发利用产生了重要影响。

1984年1月5日，国务院批准了青岛市城市总体规划，并在批复中要求把青岛市建成以轻纺工业为主，经济繁荣、环境优美、科研文教事业发达的社会主义现代化的风景游览和港口城市。这一批复对青岛的城市性质、城市功能作出了定位，胶州湾的发展前景得以确立。

1985年，青岛海洋资源研究开发保护委员会成立，组织各方面专家开展了对胶州湾的研究工作。其一系列研究及其取得的成果，为后来胶州湾及邻近海岸带功能区划创造了有利条件。

1994年，胶州湾规划、保护联席会议成立，并会同相关部门和单位组织，对环胶州湾进行了全面调查。在此基础上编写了《胶州湾及邻近海岸带功能区划》，为合理开发、保护胶州湾提供了科学依据。

1995年5月通过的《青岛近岸海域环境保护规定》、《青岛海岸带规划管理规定》两部法规，则将开发、保护胶州湾正式纳入法制轨道。

2004年下半年，青岛市人民代表大会城建环保委员会组织专题调研组走遍胶州湾沿岸，向所有用海单位进行调查，登门拜访征求了海洋环境专家的意见，与市政府及其有关部门进行了多次沟通，在此基础上起草了一份保护胶州湾的议案。2005年1月，市第十三届人大三次会议讨论通过了《关于切实加强胶州湾水域及近海岸线保护的议案》。这意味着，对胶州湾的保护工作已提升到新的战略高度。

七、建议及对策

尽管政府已经警觉胶州湾面临的境况，并作出了相应的政策反应，但笔者认为还应加强以下几方面的管理措施。

（1）应该强调从生态系统出发，用生态的观点，进行胶州湾滩涂资源现状与动态调查。近些年胶州湾的滩涂资源状况变化很大，有必要对滩涂资源状况进行全面的调查，包括滩涂的空间分布与数量变化，滩涂的开发利用类型，滩涂的自然环境状况，动植物种类以及沿海地区的社会经济情况等。开展科学研究，对胶州湾区域进行长期监测，积累科研资料并掌握其发展演化规律。根据这些调查资料来判断滩涂资源开发的合理性，从生态角度上作适当的开发利用规划，实现滩涂资源的可持续利用。

（2）充分做好围填海工程的环境影响评价工作。滩涂资源的开发应根据海洋环境容量，确定开发的限度，把合理利用和资源保护结合起来，尽可能减少对生态环境的影响。虽然大多数围填海工程项目依照有关法规都做过环境影响评价，但是每一个工程的环境影响评价基本上只是分析本项目对周围生态环境可能造成的影响，未考虑湾内其他已建围垦项目对生态环境的综合叠加影响。今后滩涂围垦工程的环境影响评价工作应从单一的项目环境影响评价转移到兼顾多项目环境影响综合评价上来。

（3）编制详细的围海造地项目指南和围海造地管理规划；完善论证与评

审制度，包括可行性报告（工程可行性、投入产出分析、内部收益率）、环境影响评价报告（自然资源价值评估、水质状况、生态系统安全、侵蚀与淤积等）、审批制度及公众参与机制；制订围填海项目跟踪监测、后评估制度；建立中长期监测及社会经济评价机制。

（4）规范海域使用执法监督检查制度，强化海域使用管理的执法工作。青岛市人民政府海洋行政主管部门及其所属的中国海监机构要加大执法力度，整顿和规范海域使用管理秩序，对《海域法》实施后未经批准非法占用海域，越权批准或者不按海洋功能区划批准使用海域，擅自改变海域用途等违法行为，要按照《海域法》的有关规定追究有关当事人的法律责任。

（5）严格执行青岛市第十三届人代会提出的六条建议。①有效控制填海。②控制滩涂养殖围垦，对已经失去经济价值的虾池、鱼塘、盐田等要适度"退池还海"。③严格控制湾内工程设施的建设，避免造成胶州湾的淤积。④严格控制占用海域和海岸线的项目建设。在胶州湾海域及其海岸线，除法律允许并经青岛市政府批准建设的重大项目和国家重点建设项目外，暂停其他填海工程项目建设。⑤对无涉海需要的项目要另行选址，有涉海需要的要优先保证港口、码头需要，确需填海的，要严格控制填海规模，防止胶州湾面积急剧缩小是重中之重。⑥加快胶州湾污染治理步伐，采取有力措施，保护好胶州湾水域和近海岸线。因为，在一定程度上，污染可以治理，可是一旦填海造地，将永远无法还原。

（6）在编制《环胶州湾近海岸线保护和利用的控制性详细规划》的同时，加大社会宣传力度，组织相关的科技、科普宣传与政府宣传相结合，使更多的人了解胶州湾在经济、环境和文化方面的重要价值；正确处理短期利益与长远发展的关系，逐渐增强保护胶州湾的社会意识。在科技宣传、社会舆论和法律的共同约束下，开展胶州湾的保护和恢复工作。

参考文献

[1] [8] [9] 张维迎. 博弈论与信息经济学 [M]. 上海：上海人民出版社，2004：2, 8, 48-50, 65-66.

[2] 刘洪滨. 胶州湾成因的探讨 [J]. 海洋地质与第四纪地质. 1986, 6 (3)：53-65.

[3] 郑全安，吴隆业. 胶州湾遥感研究——总水域面积和岸线长度量算 [J]. 海洋与湖沼，1991, 22 (3)：193-199.

[4] 贾怡然. 填海造地对胶州湾环境容量的影响研究 [D]. 青岛：中国海洋大学，2006.

［5］薛荣俊，隋波，薛巍. 海洋走航式单边连续观测地震折射波 $\triangle t$ 解释方法——一种新的浅层折射单支时距曲线解释方法［J］. 中国海洋大学学报，2004，34（2）：269，272.
［6］邹卫东. 保护好"母亲湾"［N］. 人民代表报，2005 – 07 – 02.
［7］洪光，刘春光. 青岛市气候变暖的特征［J］. 气象，1997，23（8）：57.

A Game Analysis of the Reclamation Behavior and Research on Protecting Strategy

Sun Li, Liu Hongbin

[Abstract] Over the last 10 years, there were large areas of reclamation from the coast around Jiaozhou Bay. Consequently, some severe and negative impacts have been caused on the coastal environment, which is shown clearly in the following aspects. It caused the dramatic changes of coast and seabed landform; the amount of tidal water and environmental capacity decreased; water quality has been deteriorated. Thus, the biodiversity is decreasing, and more and more plants and animals are disappearing.

In this article, static game models of different behavior between some stakeholders have been analyzed. Based on depicting the environmental impacts which caused by these actions and looking back measures which had been taken, strategies and suggestions for controlling reclamation in Jiaozhou Bay are put up, in order to support the sustainable development in the coastal zone.

[Key Words] Jiaozhou Bay Game Reclamation Protection

JEL Classification: Q25, R52, C73

国际及区域创新体系建设：理论进展与海洋创新体系实证[*]

刘曙光　朱翠玲[**]

【摘要】 在对国家及区域创新体系理论文献回顾的基础上，本文对其国际动态研究进行了综述，从国家及区域创新体系的框架中探索海洋创新活动的规律，概述了海洋创新体系的主要内容，包括创新平台和产业集群建设，并且指出其现有的问题，旨在对尚不完善的海洋创新体系建设提供借鉴。

【关键词】 国家创新体系　区域创新体系　国际动态　海洋创新体系

一、导论

国家创新体系是20世纪80年代以来为适应知识经济时代的到来，提升国家整体竞争力而提出的一个重要学术概念，此后不断得到发展和深入，而且关于分部门/行业、分地区的国家创新体系建设的研究越来越多。随着全球化、信息化、市场化的发展，经济的区域化与个性化的趋势十分强劲，资源和分工

---

[*] 教育部人文社会科学重点研究基地重大项目"国家海洋创新体系建设的战略组织研究（07JJD630012）"初期成果。

[**] 刘曙光，博士，教育部人文社科重点研究基地中国海洋大学海洋发展研究院、中国海洋大学经济学院教授，青岛：266005，Email: dawnliu9631@263.net；朱翠玲，中国海洋大学经济学院硕士研究生，青岛：266071。

在不同层次上进行整合，并且越来越集聚于有个性的地区。这种区域化的发展趋势也使人们的研究视野开始转移，如何实现区域经济的协调有序发展成为理论研究和实践的重要领域。传统的区域经济单纯依靠外来资本及本地资源的传统增长方式已经不适应经济发展的要求。如何实现区域经济的协调和可持续发展已经成为区域经济研究的重要任务。因此，将创新系统理论引入区域经济理论之中，以创新来带动区域经济发展的区域创新理论的兴起，是区域经济发展理论研究的必然结果。而随着全球资源短缺和环境恶化现象的凸显，海洋已经成为各国竞争的焦点，海洋经济既是水体资源经济、海产业经济，又是区域海洋经济，海洋创新体系是国家创新体系建设的组成部分，建设海洋创新体系是提高海洋开发能力和水平、加快海洋经济发展的需要。本文通过追溯国家创新体系、区域创新体系概念的来源，着重对其近几年的研究动态进行综述，概述了海洋创新体系的主要内容，包括创新平台和产业集群建设，并且指出其现有的问题，旨在对尚不完善的海洋创新体系建设提供借鉴。

二、国家创新体系进展

美籍奥地利经济学家约瑟夫·阿罗斯·熊彼特（1912）第一次将创新（innovation）概念引入经济学。继熊彼特以后，不少经济学家对熊彼特的创新概念进行了深化研究，英国学者弗里曼（Freeman，1987）在其著作《技术和经济运行：来自日本的经验》中，首次提出国家创新体系（National Innovation System，NIS）这一概念。他把 NIS 定义为"一种公共和私营部门的机构的网状结构，这些公共和私营部门的行为和相互作用创造、引入、改进和扩散新技术。"1992 年，他又进一步把国家创新系统分为广义和狭义两种。广义的国家创新系统包括国民经济涉及引入、扩散新产品的过程和系统的所有机构；狭义的国家创新系统涵盖了与科技活动直接相关的机构，包括大学实验室、产业的研究开发实验室、质量控制和检验、国家标准机构、国立研究机构和图书馆、科技协会和出版网络，以及支撑上述机构的、由教育系统和技术培训系统提供的高素质人才。纳尔逊（Nelson，1993）对不同的国家创新体系作了一个详细的比较，将国家创新体系定义为"一系列的制度框架，他们的相互作用决定着一国企业的创新能力（业绩）"，因此，国家创新体系是一种将制度安排与一国的技术经济实绩相联系的分析框架。梅特卡夫（1994）也认为："国家创新体系是一套明确的制度，这套制度可以共同或单独行使，有助于新技术的开发和传播，另外，这套制度为政府制定和执行影响创新进程的政策提供框架。这是一套相互联接的制度体系,用于知识、技能和新技术产品的创造、储存和转让"。OECD（1997）认为,国家创新体系是"公共和私人部门中的组织结构网络,这些部门的活动和相互作用决定着一个国家扩散知识和技术的能力,并影响着国家的创新业绩"。自2002年起,其关注国家创新体系的内容包括:①创新型企业识别、定义、特征描述及其培育;②不同知识生产者（企业、大学、科研院所）的知识创新,以及相互知识流动和耦合;③公共部门对知识创新的支撑体系和协同;④国家体系的国际化和空间外延性。

在近些年的研究中,更侧重于对国家创新体系的运行机制的研究。基本遵循"知识和信息流动为核心"（OECD,1997）,这种流动随着研究的不同侧重形成了非常清晰的研究层面,不同层面的研究往往互相交叉、相互渗透,中介组织/机构（agents）在创新结网和创新维持中的作用至关重要,同时各种非正式组织的指示传播为创新系统提供了给养,对于这种现象,斯多波和威那波尔用"言传"（buzz）来比喻（Storper and Venables, 2002）。"言传"形容人们或企业由于在同一产业、同一地方或区域里共存或共生,通过面对面交流所形成的信息和通讯生态。同时也有越来越多的文献直接涉及国家创新体系的国际战略合作,如基于东亚国家和地区经验的战略合作研究的成效和教训分析（Sakakibara and Dodgson, 2003）,爱尔兰如何借助欧盟的创新联合来发展本国知识体系（Kostiainen and Sotarauta, 2003）,欧盟与中东和北非的技术转移和国际创新合作（Koehler and Wurzel, 2003）,以及近来欧洲学者关注的如何通过国家创新体系的力量推进和提升欧洲中小企业（群）在国际化过程中的创新能力（Berger and Hofer, 2007）。

国家创新体系理论的研究基本上是在国家技术创新系统理论成果的基础上,通过对技术创新过程的整体性思考,并运用系统的理论与方法发展而来的。目前,国家创新体系尚没有统一定义,国家创新体系研究也没有形成完整的理论体系、共同的学术规范和适用边界。但国家创新体系理论的本质内涵只有一个,即科学技术知识在国民经济体系中的循环流转及其应用,而且在经济全球化的大背景下更关注于国际间知识的流动和技术的战略合作。

三、区域创新体系进展

20世纪80年代后期,在创新研究领域内,出现了与技术创新和制度创新不同的创新理论,即从系统的观点来研究创新的新思路,提出了创新系统理论。90年代以来，以城市与区域规划专业为主的专家学者在参与城市与区域的开发和管理及国家创新系统研究过程中，关注并研究创新系统建设与区域的密切关系，波特（Porter，1990）将区域创新系统看做是国际创新系统的一部分。其中英国卡迪夫大学的库克教授（Cooke，1990、1994、1998）最早对区域创新系统进行了比较全面、系统的研究。综观20世纪90年代以来区域创新系统的研究状况可知，区域创新系统的概念应包括以下基本内涵：具有一定的地域空间范围和开放的边界；以生产企业、研究与开发机构、高等院校、地方政府机构和服务机构为创新主要单元；不同创新单位之间通过关联，构成创新系统的组织结构和空间结构；创新单元通过创新（组织和空间）结构自身组织及其与环境的相互作用而实现创新功能，并对区域社会、经济、生态产生影响；通过与环境的作用和系统自组织作用维持创新的运行和实现创新的持续发展。研究的关注点主要有区域创新系统环境研究，区域创新系统组织结构研究，区域创新系统空间结构研究，区域创新系统功能和过程研究等。

自欧文－史密斯（Owen-Smith）和鲍威尔（Powell，2002）在波士顿生物技术界的案例中使用"pipelines"来描述远距离的、非本地的信息相互结合的渠道以来，全球性渠道（global pipelines）的重要性越来越多地受到重视。欧文－史密斯和鲍威尔指出新知识的传输不仅仅是通过区域内部的相互接触，而是更多地依靠全球性的渠道和地方互动的形式。贝谢尔特、马尔姆博格（Bathelt，Malmberg）和马斯科尔（Maskell，2004）提出地方互动，同时指出"地方互动（local buzz）"和"全球渠道"之间的对应关系，即地方信息流、交流不仅仅是通过地方内部互动的方式，而是通过地方互动和全球渠道这种强化的方式。并且进一步指出集群中的企业所建立跨区域的信息流通渠道越多，流到区域内的市场和技术信息就越多。全球性渠道可以加强企业的凝聚力，有利于空间知识、创新的产生，进而不同类型知识的结合及信息的传输才能更好地解决科技、组织、商业等问题。强调不要忽视全球创新联系对于地方（国家）创新的重要性（图1）。

日本学者馆尾有本（Tateo Arimoto，2006）从投入—产出的角度分析了创新的过程，列出了互动系统。在创新生态系统内部，知识创造和长期的基础研究作为创新的投入，在生态系统内与人力、资金相结合，形成良好的区域集群、产学协作等机制（包括管理机制和评价的标准），最后产出新产品/服务、开拓新市场、提供社会服务，从而增加企业盈利和社会福利。其中资金支持、人力资源培训以及公众的可接受性保障着创新生态系统的有序运行（图2）。

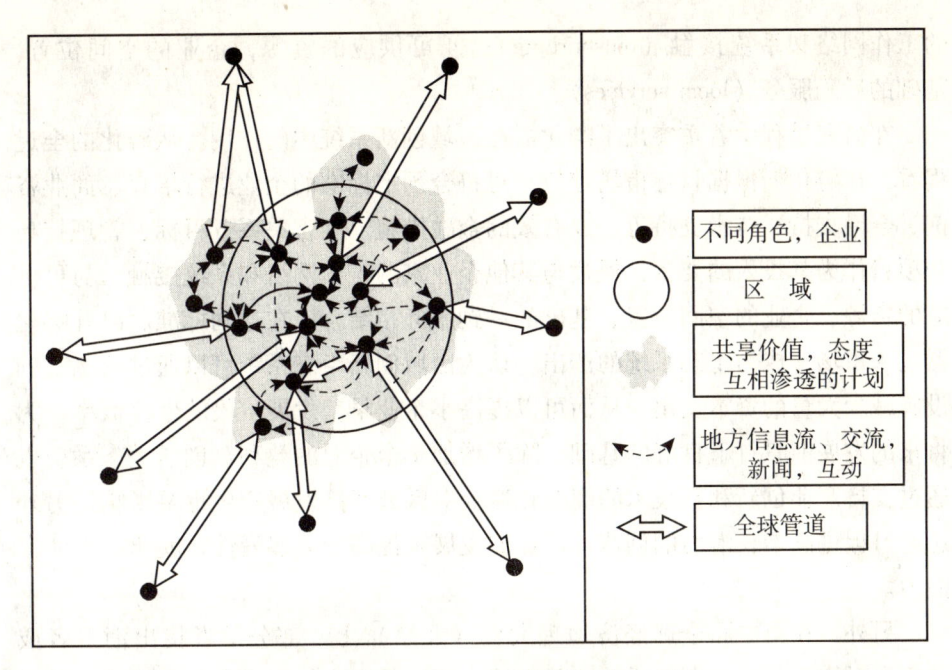

图1 区域创新的全球性渠道与地方互动

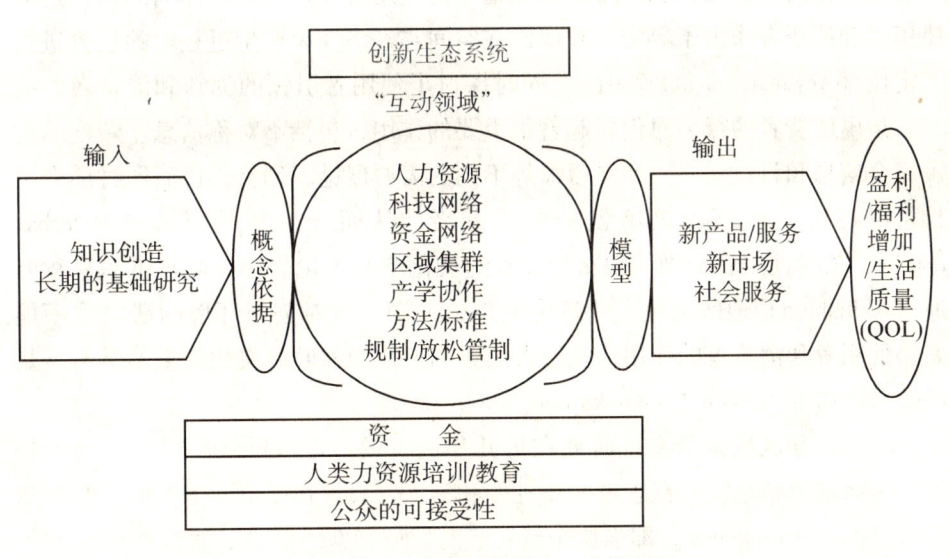

图2 创新生态系统的投入—产出分析

从同样的角度出发阐释各个因素的作用机制，欧洲跨学科研究所（EIIR，2006）认为创新网络生态系统的成功主要是归结为以下基础因素的有效结合：①与商业有关的经历、技术、知识和专家意见（expertise）；②在商业管理、市场营销、技术和金融领域的企业网络（chain and cluster）；③与其他企业家的工作网络和系统接触（outsourcing）；④可供应的资源，企业的空间位置，基础的后勤服务（local services）。

在研究过程中着重突出了孵化器在区域创新系统中的作用。从孵化的全过程看，在孵化前根据目标市场定位，对符合预定标准的企业进行培育、商业咨询、金融支持、技术支持等，以有效的金融措施、利害关系方目标、管理技巧和项目作为其投入的要素，通过与其他企业家的工作网络和系统接触，与可供应的资源、企业的空间位置、基础的后勤服务相结合，在达到标准后离开孵化器进入市场，成为生态系统的产出。认为区域创新生态系统可以通过共享基础设施减少大量的资本支出，从而可以提高小企业早期发展阶段的生存概率，为将来的发展前景打下良好的基础；除了增加单个企业的盈利，创新生态系统也通过支持产业创新和新技术的商业化途径对区域经济发展产生重要影响；另外还通过促进高增长潜力的创造性产业的发展来提高产业多元性、就业和企业家能力。

另外，在第二届全球经济地理大会（北京）上，部分学者指出消费者或使用者对知识创造和创新活动的关键性作用，关注诸如健康产业、游戏、动漫产业等，特别是在知识和创新产出模型中的使用者如何进行合作的问题，还对使用者和消费者作出了界定，对他们在空间和经济知识产生过程中的位置进行了定位（Aoyama, et al, 2007）。同时探讨了使用者引领的创新和消费者的能力，指出消费者的行为对创新起着很重要的作用，包括不对称信息、新产品的不完全信息和科技路线、消费习惯等不同的需要促进了创新，进而通过中介机构可以将需求信息反馈给创新的发起者，从而产生新的科技（Malerba, 2007）。在《使用者引领的创新与日本的文化产业》论文中，以日本东京秋叶原（Akihabara）地区文化产业群为案例，强调了在对创新环境的研究中不能忽略使用者和消费者的作用，使用者引领创新作为秋叶原文化产生的基石，是联接公司内部的纽带（Nobuoka and Jakob, 2007）。

综观国际区域创新系统研究在近几年的发展，可以反映出以下方面的特征：①区域创新系统的概念进一步延伸和深入，表现在以下三个方面：全球治理（global governance），即强调全球背景对区域创新的影响。通过外部（国外/区外）技术扩散，促进各种机构、组织及代理机构的互动；单元间组织学习（organizational learning）和个人间沟通（面对面交流，tacit communication）；专业中介组织（agents）和公共基础性平台（综合性孵化器、虚拟网络平台、论坛等）；②强调区域创新系统的生态化与社会可持续发展相结合，注重区域创新系统中各因素之间的整合利用；③倾向于使用者引领的创新，注重消费者和生产者对知识创造和创新的关键作用。

四、海洋创新体系实证研究进展

(一) 海洋创新体系概念的讨论

加拿大政府十分重视国家层面海洋创新体系建设,认为:海洋创新体系远比其作为一个传统的单一产业复杂得多,它不仅包括产业部门或技术层次的要素(船只建造、水产养殖业、海洋电子),还包括环境治理、信息获取、规章制度和具体机构,以及对知识需求的反馈。这种创新体系需要持续的关于自然环境的知识和信息流动;涉及经济行为者的知识路径发展;更大范围司法管辖区和产业活动的合作;建立对非商业性和商业性海洋环境敏感的政府机制;基于研究、教育、拓展和协调一致的活动建立的海洋创新体系机构也必须反映这种多样性的关系和行动者(Conseil national de recherches Canada, 2005)。澳大利亚的南澳州国家海洋创新体系的目的是使其成为世界级的海洋科学、教育和产业发展中心,其研究集中于四个重要领域:海洋食品质量和价值增值,生态系统服务,水产养殖业创新,生物安全。[①]

瑞典的区域海洋创新体系建设,突出了在构成船只所有者的网络中(银行、航运代理等)经营者之间紧密合作的重要性,也强调了港口的重要性,海洋中心的发展大部分设在较大的港口地区,在瑞典的案例中也显示出港口的发展很大程度上影响航运地区的发展,分析显示港口经营组织在海洋创新体系中起着重要的作用。其发展取决于地方海洋产业群之间的竞争,大区域产业群之间更复杂的竞争可以创造良好的创新环境(Palmberg, Johansson and Karlsson, 2006)。海洋产业在挪威的海洋技术发展中也起着很重要的作用,其海洋创新体系由三部分构成:技术供给者、科研机构和农业/渔业企业。传统水产养殖业创新体系显示直接供给者在产业和科研机构、其他知识传输服务和政府管理之间起着纽带的作用,供给者被看做是研究成果的购买者,企业可以自发地进行一些科研活动而不是通过供给者购买科研成果,但现在这种创新体系发生了变化,在产业结构中更依赖大的公司,更多的研究服务被产业购买。认为处于核心地位的是农业/渔业企业,其次是技术提供者,再次是科研机构(Olafsen and Sandberg, 2007)。

---

① 详见:http://www.sardi.sa.gov.au。

（二）海洋创新体系建设的主要内容

为了顺应国家发展不同科技革命背景下的国家战略需求，适应海洋资源勘探开发和海洋领域的产学研一体化建设，以及信息技术、新材料技术和生物技术等在海洋研究中的渗透，自18世纪初，各发达国家陆续建立各种国家海洋研究中心，开始有计划、有步骤地进行海洋科学基础研究和应用研究的探索，形成了目前世界几所主要的海洋中心：美国拥有两个世界著名的海洋科研机构——斯克利普斯海洋学院和伍兹霍尔海洋研究院，瑞典的可瑞斯堡海洋研究站，俄罗斯科学院P. P. 希尔绍夫海洋资源研究院，日本海洋－地球科学技术中心，澳大利亚海洋科学研究院，法国海洋开发研究院，英国南安普顿国家海洋中心。这八家海洋中心均由政府或国家科学院倡导成立，为公益性非营利组织，基于海洋问题的高度复杂性、国际性和高投入成本，在其运作过程中的资金投入和支持有四种情况：高等院校投入，联邦机构联合投入，大学与科研院所的联合投入，政府拨款。其科研目的主要是进行海洋领域基础知识的研究，通过向政府及其他组织提供年度报告和政策建议，向公众开放海洋图书馆等方式，带动国家乃至全球对海洋科学的重视和研究。海洋应用研究的成果多以商业项目出售转让的方式为主，通过企业来实现研究成果的产业化（刘曙光、于谨凯，2006）。

海洋创新体系的建设对于产业集群的发展升级起着重要的作用。海洋产业集群的发展很大程度上取决于领军企业与相关中小企业的创新关联，需要国家和地方政府的支持，表现在协助创建产业集群中介组织，使产学研相结合；已有的海洋产业（集群）还没有真正形成与国家海洋研究中心建设一样对等的国家级海洋创新体系子系统，而这也是各国正在尝试的重要方向（刘曙光，2007）。

（三）海洋创新体系建设中存在的问题

综上所述，可以发现：①对于海洋创新体系建设问题，还缺乏严格意义上的直接论述文献，没有将个案和局部系统分析的结论上升到一般结论之中，目前尚没有形成完善的海洋创新体系，缺乏对整个体系框架的设计。②由于海洋经济的复杂性，其创新系统是复杂的网络系统。主要由海洋知识创新系统、海洋技术创新系统、海洋创新管理系统、海洋创新服务等系统组成，而海洋经济创新的本质即创新系统的整合和创新过程的协同。但是现有的文献对海洋创新体系系统间的研究存在着不足，没有划分和确定各子系统及相关要素，也没有确定子系统间关系的因子权重和关联程度。

五、结论与启示

本文对国家创新体系、区域创新体系以及海洋创新体系的相关理论及国际动态进行了综述，其中：①在国家创新体系层面，近几年更多的研究侧重于对国家创新体系运行机制的探讨，强调知识和信息流动对国家创新体系建设的重要作用，同时也有越来越多的文献直接涉及国家创新体系的国际战略合作，关注在经济全球化背景下通过国际间技术和知识的转移而进行的跨国合作，从而促进国家创新体系建设。强调国家创新体系的存在决定了一国技术创新能力，国家在创新过程中担负领导角色，国家是创新活动的决定因素。②在区域创新体系层面，主要探讨其在信息流动的全球性渠道和地方互动的形式；与社会可持续发展相结合，强调区域创新系统的生态化；注重消费者和生产者对知识创造和创新的关键作用，倾向于使用者引领的创新。

以上通过对国家创新体系和区域创新体系近几年新进展的探讨，为海洋创新体系的建设提供理论基础，在海洋创新体系建设层面，通过总结世界主要海洋中心的建设经验，以及海洋产业集群的理论和发展升级的实证，得到以下两点启示：第一，在进行海洋创新体系建设时，可借鉴国家和区域创新体系的最新理论进展，更多地关注创新体系的动态研究和创新生态系统建设，提炼出相应的建设模式。第二，可将国家级海洋研究中心、海洋产业集群的建设和运作作为案例，跟踪研究其发展态势，重视其运作的网络系统，分析其可能涉及的有关海洋创新体系组成的子系统和要素，找出各子系统及要素所起的作用及其关联性，并且通过阶段性成果的应用来推进海洋创新活动的升级。

参考文献

[1] Aoyama, Yuko and Power, Dominic. User-led innovation, knowledge and economic geography. Second Global Conference on Economic Geography, 2007.

[2] Bernard'l Cabrer-Borr'as, Guadalupe Serrano-Domingo. Innovation and R&D spillover effects in Spanish regions: A spatial approach. International Journal of Industrial Organizations, 2006: pp. 1357 – 1371.

[3] Conseil national de recherches Canada. Technology Roadmap, 2005.

[4] Cooke P, Uranga MG, Etxebarria G. Regional systems of innovation: an evo-

lutionary perspective. Environment and Planning A, 1998, 30 (15): pp. 63 – 84.

[5] Cooke P. Systemic Innovation: Triple Helix, Scalar Envelopes, or Regional Knowledge Capabilities, an Overview. International Conference on Regionalisation of Innovation Policy-Options &Experiences, Berlin, June 4 – 5, 2004.

[6] Cooke P. The role of research in regional innovation systems: New models meeting knowledge economy demands. International Journal of Technology Management, 2004, 28 (3 – 6): pp. 507 – 533.

[7] David Doloreuxa, Saeed Parto. Regional innovation systems: Current discourse and unresolved issues. Technology in Society, 2005 (27): pp. 133 – 153.

[8] Davies, Tamsin. Developing innovation in a peripheral region: university-industry links in Wales. Second Global Conference on Economic Geography, 2007.

[9] Dosi G, Freeman C, Nelson R, Silverberg G, Soete L. Technical change and economic theory. London: Pinter, 1998.

[10] EIIR. PRIVATE SECTOR-LED ECONOMIC DEVELOPMENT AND COMPETITIVENESS BUILDING STRATEGIES: Practices from Regional Economic Ecosystems, 2006.

[11] Frank Moulaert and Farid Skeia, Territorial Innovation Models: A Critical Survey, Regional Studies, 2003 (37), pp. 289 – 302, .

[12] Fritsch, Michael and Slavtchev, Viktor. What Determines the Efficiency of Regional Innovation Systems? Jena Economic Research Paper No. 2007 – 006, 2007.

[13] Graedel T. E, Allenby B R. Industrial Ecology. Englewood Cliffs: Prentice Hall, 1995.

[14] Harald Bathelt, Anders Malmberg and Peter Maskell. Cluster and Knowledge: Local Buzz, Global Pipelines and the Process of Knowledge Creation. Danish Research Unit for Industrial Dynamics Druid Working Paper, No 02 – 12, 2007.

[15] Karlsson C, AnderssonÅE. Paul Cheshire and Roger R Stough, Innovation, dynamic regions and regional dynamics, Electronic Working Paper Series Paper No. 89, 2007.

[16] Koehler S., Wurzel U. G. From Transnational R&D Co-operation to Regional Economic Co-operation: EU-Style Technology Policies in the MENA Re-

gion. Mediterranean Politics, 2003, 8 (1): pp. 83 – 112.

[17] Kostiainen S., and Sotarauta P. Building a Knowledge Economy in Ireland Through European Research Network. European Planning Studies, 2003, 11 (4): pp. 395 – 413.

[18] Malerba, Brusoni. Perspectives on Innovation. Cambridge University Press, Cambridge, 2006.

[19] Mikel Buesa, Joost Heijs, Mónica Martinez Pellitero and Thomas Baumert. Measuring Scientific and Technological Progress. Eighth International Conference on Science and Technology Indicators, 2004.

[20] Nelson R, Winter S. An evolutionary theory of economic change. Cambridge: Cambridge University Press, 1982.

[21] Nobuoka, Jakob. User-led innovation and Japanese culture industries. Second Global Conference on Economic Geography, 2007.

[22] Olafsen T, Sandberg M G. The technological marine industry in Mid-Norway. SINTEF Fisheries and Aquaculture International Projects and Consulting, 2007.

[23] Oughton C. The regional paradox: innovation policy and industrial policy [J]. The Journal of Technology Transfer, 2002 (27). pp. 97 – 110.

[24] Owen-Smith, J. and Powell, W. W. Knowledge Networks in the Boston Biotechnology Community. Paper presented at the Conference on "Science as an Institution and the Institutions of Science" in Siena, 2002.

[25] Palmberg J, Johansson B, Karlsson C. Den svenska sjöfartsnäringens ekonomiska och geografiska nätverk och kluster, maj Paper, No 34 – 39, 2006.

[26] Rantisi, Norma. Local innovative dynamics and the global fashion city phenomenon: Montréal's missing link. Second Global Conference on Economic Geography, 2007.

[27] Sakakibara M., Dodgson M. Strategic Research Partnerships: Empirical from Asia. Technology Analysis & Strategic Management, 2003, 15 (2): pp. 227 – 245.

[28] Schuldt, Nina1 and Bathelt, Harald. Temporary face-to-face contact and the ecology of global buzz. Second Global Conference on Economic Geography, 2007.

[29] Storper, M. and Venables, A. J.: Buzz: The Economic Force of the City. Paper presented at the DRUID Summer Conference on 'Industrial Dynamics of

the New and Old Economy—Who is Embracing Whom? 2002.

[30] Tateo Arimoto. Global Innovation Ecosystem. International Conference on Science and Technology for Sustainability, 2006.

[31] 刘曙光,徐树建. 国际区域创新系统研究进展综述 [J]. 中国科技论坛, 2003, (3).

[32] 刘曙光,于谨凯. 海洋产业经济前沿问题探索 [M]. 北京:经济科学出版社, 2006.

[33] 刘曙光. 海洋产业经济国际研究进展 [J]. 产业经济评论, 2007, (1): 170-190.

[34] 汤尚颖,曹勇涛,程胜. 区域形态创新模式与区域发展 [J]. 理论探索, 2007 (5): 79-83.

National and Regional Innovation System: Theoretical Approaches and Empirical Studies of Marine Innovation System

Liu Shuguang, Zhu Cuiling

【Abstract】 Based on the review of the literatures on national and regional innovation system, the paper summarized the main international dynamics. The authors of the paper attempts to explore the marine innovation activities from the framework of the national and regional innovation system. The paper listed main contents of marine innovation system as innovation platform and the construction of industry cluster. Finally we pointed out the existing problems and aimed at improving the innovation system of marine construction.

【Key Words】 National innovation system　Regional innovation system　International dynamics　Marine innovation system

JEL Classification: R53　O31　Q58

中国海洋渔业产业化发展模式探讨

邵桂兰　李洪铉　张　希[*]

【摘要】海洋渔业产业化对新阶段海洋渔业发展战略的转变具有重要意义。现阶段，中国海洋渔业产业化的发展模式主要呈现龙头企业带动型、主导产业带动型、市场带动型、中介组织带动型四种模式。随着生产力的发展，需要不断创新海洋渔业产业化发展模式，其中"五合一"模式、"渔户+专业合作组织+企业"模式、"超市+水产品加工企业+渔户"模式将以更加成熟的运作方式成为中国海洋渔业产业化发展的新模式。

【关键词】海洋渔业产业化　发展模式　创新

一、引言

继承包制、规模经营等热点问题之后，海洋渔业产业化近十年来又成为中国海洋渔业发展中新的热点，已引起海洋渔业界的广泛关注。海洋渔业产业化的基本内涵是指以市场为导向，以加工企业为依托，以广大渔户为基础，以科技服务为手段，通过把海洋渔业生产过程的产前、产中、产后诸环节联接为一个完整的产业系统，实现捕养加、产供销、渔工贸一体化经营。海洋渔业产业化对解决目前海洋渔业市场经济中所遇到的困难，促进海洋渔业和渔村经济实

---

[*] 邵桂兰，博士，中国海洋大学经济学院教授，Email: shaoguilan@public.ad.sd.cn；李洪铉，博士，中国海洋大学经济学院硕士研究生；张希，中国海洋大学经济学院硕士研究生；青岛：266071。

现两个根本转变,解决渔村改革的深层次问题,实现海洋渔业自我发展、自我积累、自我约束、自我调节的良性循环,建设海洋渔业现代化具有重大意义。

目前中国的海洋渔业产业化已形成以下主要特点:一是产前、产中和产后各个环节有机地联接了起来,形成了完整的产业系统和产业链;二是通过龙头企业的带动,把分散的家庭小生产纳入了社会化大生产的轨道,提高了海洋渔业生产的社会化程度和组织化程度;三是小规模经营与大市场的联接,促进了生产要素的优化组合和合理配置,使产业结构和产品结构不断优化;四是推动了海洋渔业的体制改革,各渔区努力尝试运用现代企业的管理办法组织海洋渔业生产经营,建立了合理的利益分配机制。然而由于仍处于产业化经营的初期发展阶段,中国与发达国家相比还有一定的差距,还存在着渔农业户小规模、分散经营,海洋渔业龙头企业偏少,海洋渔业建设支撑体系滞后,水产加工总体水平不高等问题。经过十多年的探索,在海洋渔业产业化的实践基础上,中国渔村已经摸索出了多种海洋渔业产业化模式。为了在全国范围科学、有效和稳步地推进海洋渔业产业化,有必要对现有的海洋渔业产业化模式加以分析和探讨,在此基础上进一步探索适合中国国情的海洋渔业产业化发展的新模式。

二、中国海洋渔业产业化发展模式

中国海洋渔业产业化发展的模式,从产业链驱动力的角度,可分为龙头企业带动型、主导产业带动型、市场带动型、中介组织带动型、农民组织带动型、公司+农户型、公司制农业型和科技带动型等。按照联合的紧密程度,可分为松散型、半紧密型和紧密型三类。下面我们分别对几种主要模式进行分析。

(一)龙头企业带动型

这种模式是以水产品加工、冷藏、运销企业为龙头,围绕一项产业或产品,形成"公司+基地+农户"的产加销一体化经营,采用契约、股份制等形式与渔户形成利益联接,分为龙头加工企业带动型和龙头经销企业带动型。

推进海洋渔业产业化,引导渔民进市场,龙头企业的牵动能力大小直接影响着海洋渔业产业化经营的规模和成效。因为龙头企业是海洋渔业产业化链条上与市场对接最紧密而又能起主导作用的一个重要环节,是上联市场下联农户的枢纽。它可以根据市场需要配置海洋渔业资源,引导生产要素合理流动;还可以将渔民产品集中起来，通过多层次加工，实现大幅度增值，甚至变废为宝，不仅从量上形成了较大的商品规模，而且从质上改变了水产品的原始形态，适应了日益提高的消费水平。以龙头企业为载体打入国内外市场，通过储藏、保鲜、加工转换，可调节需求品种的变化和需求层次的变化，提高水产品的市场竞争力，还可形成可观的批量和持续输出量，使之具有更大的市场占有率。更重要的是龙头企业具有促进海洋渔业技术的普及和提高的作用，它可以通过加工这个环节，采用新工艺流程、新技术，提高产品科技含量，提高附加值；通过科技服务的功能，解决技术难题，从而使渔户小生产跃上一个新台阶。这种模式要求水产品不易变质、便于储运、加工深度大、增值率高；要求龙头企业有一定的资金投入、一定的技术、一定的规模；还要求经销龙头企业有丰富的市场经验，有进入国内外市场的良好渠道。它只适用于经济较发达地区和交通便利、易吸收外资的海洋渔业区域。

这种模式是目前中国海洋渔业产业化最重要的发展模式，也是带动渔民增收的基本模式，已取得了较好的经济和社会效益。以江苏省为例，截至2006年上半年，全省水产品加工企业达753家，水产品加工总量达58.3吨，产值超过80亿元，其中绝大多数企业都采取这一模式带动渔民增收。但是这一模式也暴露出一些潜在的问题。由于龙头企业与渔民都是经济体，在市场经济条件下，都追求利益最大化，一旦水产品市场价格有较大变动，企业与渔户之间签订的合同就往往难以落实。同时，企业强市、渔户弱势的状况没有明显改变。由于龙头企业可以为地方财政增加税收，因此当龙头企业与渔户发生矛盾时，不少地方政府大都保护龙头企业。于是，大多数渔户难以真正获得水产品加工增值的利益，根本原因在于大多数的龙头企业不是渔民自己组织的合作制企业。

（二）主导产业带动型

这种海洋渔业产业化模式从利用当地资源、发展传统产品入手，形成区域性主导产业。从"名、优、特、新"产品的开发入手，对那些资源优势最突出、经济优势最明显、生产优势较稳定的项目，重点培养，加快发展，形成新的支柱产业，围绕主导产业发展产加销一体化经营。

主导产业也叫主导增长产业，它是指那些能够迅速和有效地吸收创新成果，对其他产业的发展有着广泛的影响，能满足不断增长的市场需求，并由此而获得较高的和持续的发展速度的产业。这种模式有利于提高海洋渔业比较利益，改变海洋渔业弱质地位，增加海洋渔业投入，为海洋渔业经济增长方式转变提供物质基础。能够打破一、二、三产业的分工界限，渔民自己生产、加工、销售，实现了海洋渔业利润最大化。正确确立主导产业，可以在稳定传统产业的基础上，积极发展新兴产业，实现海洋渔业的产业升级，整体提高海洋渔业的价值水平。可以因地制宜，发挥区域化比较优势，适应市场需求，发展适销对路的产业和产品，改变过去一味提高资源利用量的做法，着重提高资源利用率，从而优化农业结构。

这种模式要求该地区资源独特，能大量生产各种名、特、优水产品；要求农民及其领导有较高的素质，能适应国内外市场需求，正确选择和确立主导产业，创造品牌和名牌。中国东部沿海地区的不少县区在海洋渔业领域都拥有这种区位优势。但如何选择和培育主导产业，实现传统产业的更替，却是一项难题。目前，中国主导产业在选择上是多条龙并存，在布局上是区外雷同、区内零乱，在建设上是低水平重复。如江西南丰种橘养蟹两业并举，20 世纪 90 年代中期以后养蟹就席卷了中国的大江南北，出现了价格一哄而下的激烈竞争态势，各地的这种支柱产业纷纷下马。可见，主导产业是渔业产业化的基础。主导产业不是主观确定的，而是根据当地资源优势在统筹规划的基础上确定的，它可以通过资源评估和项目评估等方法确定。

（三）市场带动型

这种模式是通过培育水产品市场，特别是专业批发市场，带动区域化生产和产加销一体化经营。专业批发市场作为经济实体，通过对海洋渔业产前、产中、产后服务，包括提供市场信息、优良苗种和渔用生产资料、生产技术服务等，可以引导所在地区的渔户按照市场需要调整产业结构，及时提供质量合格、数量足够的水产品。

通过建立形式多样的市场，为渔民产销见面提供商品交易的场所，可以达到"建一个市场，活一片经济，富一方群众"的目的。市场的发育，解决了水产品的流向问题，推动了海洋渔业的规模化生产，使海洋渔业的最终产品、各种中间产品、劳务和消费品以及各种海洋渔业生产要素进入市场，从而提高渔民的市场意识，提高水产品的商品率，提高海洋渔业商品化的程度。

该模式适合于渔户投资少、经济发展水平较低的地区，它要求有计划、有组织、有胆识、有统一市场。但多年来在计划经济体制下进行劳作的渔民，进入市场后，一时陷入了有主权、无主张的困惑。他们无门路，不知养什么；无技术，不知怎么养；无胆量，怕承担风险。由于各类市场与渔户之间在利益上的松散结合，市场价格的风险主要由农户承担，加上市场体系不完善，地方割据现象还很严重，市场容量小、专业化程度低、辐射面窄、要素市场发育滞后等，渔民很难获得较为全面和真实的信息，从而使渔户生产具有很大的盲目性。因此，地方政府必须在收集市场信息、开拓销路、规范交易规则等方面，为渔民多做排忧解难工作。

（四）中介组织带动型

这种模式是根据市场需求，以中介组织为依托，实行跨地区联合经营，充分发挥加工企业的联动效应，逐步建成市场占有率高、竞争力强、规模大、生产要素大跨度组合和集生产、加工、销售相联接的一体化企业集团。其核心是各种海洋渔业专业协会、研究会，它们是一种自我管理、自我服务、民主决策、分户经营、风险分散、互惠互利的松散型合作组织。如渔民专业合作社、供销社，各种专业技术协会、销售协会等，这些协会一般由一个或几个专业大户牵头，或依托国家科技和业务部门中介组织，它们在信息、资金、技术、销售等方面具有优势，能为渔民的产、供、销提供各种服务，也能为加工、销售企业提供服务。

国外经验证明，各种中介组织特别是海洋渔业合作社，是渔民自己联合、民主管理、团结互助的一种有效形式，在提高渔民的组织化程度、增强水产品的市场竞争力、促进政府与渔民的关系和保护渔民的利益等方面，发挥了积极的作用。这种模式在中国被广泛采用，仍以江苏省为例，目前全省共有各种渔业中介组织，包括专业合作经济组织和协会 1 275 家，渔业经纪人队伍 31 510 个，加入组织的渔户达 3 万多户，占渔户总数的 10%。

这种模式适用于技术要求比较高的水产养殖业，尤其是在推广新产品、新品种、新方法的过程中，这是一种投资低、收益高、渔户得到实惠多的好方法。但目前中国水产品初加工的较多，深加工、精加工的较少；产品一二次增值的较多，多次增值的较少，产品的科技含量较低。其发展速度快但覆盖面较低，联合的农户十分有限。其他力量介入得多，渔民自己组建得少。各种组织尚未打破区域限制，规模仍然较小，既限制了自身的发展，也未能起到很好的示范性及带动性作用。从服务的内容看，仍以低成本的技术、信息服务为主，无法充分应对加入 WTO 后国内外市场接轨中所需要解决的问题。因此，在发展这种产业化模式过程中，必须加大政府的支持，给予一定数额的开办费和运行费，并在人员和办公条件等方面给予支持。必须重视提高海洋渔业经营主体的组织化程度，降低组织成本和运行成本，培育出有很高威望、乐于奉献、维护集体利益的领路人。要打破地区封锁，推动组织联合，为水产品走出国门开拓市场。

三、中国海洋渔业产业化的新模式

综观国内外海洋渔业产业化发展的历史，尽管形式多样、模式各异，但一种产业化模式是否具有生命力，是否能够促进经济的发展，最终要看这种模式是否与所有制结构相统一，是否与经济发展状况相适应，是否与生产力水平相一致。随着生产力的发展，不断创新海洋渔业产业化发展模式，走出一条适合中国国情的产业化发展之路是现阶段研究的重点。笔者通过分析、比较近年来探索比较成功的产业化经营模式，认为以下几种模式可以作为我国今后海洋渔业产业化发展的新的方向。

(一)"五合一"的海洋渔业产业化新模式

所谓"五合一"，就是政府、龙头企业、科研院所、银行、海洋渔业养殖户五个方面共同合起手来，发展海洋养殖业的一种组织形式。其中，政府搭建平台，发挥信用优势、政策优势、基地建设优势；龙头企业发挥市场、管理、人才、资金优势，向分散的养殖户提供扶持资金，收购养殖产品；银行发挥资金优势，通过贷款支持养殖户的发展；科研院所发挥技术、信息优势，提供科技支持；养殖户发挥自身产品、设备优势承接订单和项目。

这种模式最早由大连市长海县政府、獐子岛渔业集团、大连水产学院、农业银行和 50 户养殖户于 2004 年 12 月份开始试行，通过签订养殖合同和协议，该海洋渔业集团向养殖户提供扶持资金 940 多万元用于渔民购买物资、及时取得生产技术指导以解决养殖产品的病害问题等。这一模式一经试行，便被证明取得了良好的经济效益和社会效益，当年 11 月份獐子岛渔业集团公司按合同向养殖户收购扇贝 6 700 多吨，价格比上年增长 30%，并收回了年初发放的全部资金。

以往的"公司＋农户"的简单组合形式，难以规避市场风险；"五合一"模式加固、加长了合作"链条"，增加了防风险机制，较好地解决了龙头企业与渔民养殖业户之间的产业利益纠葛，增加了互利互惠、和谐发展的实质性内容，扩大了合作、共赢的发展空间。通过充分利用海域资源，积极发挥各方的优势企业迅速做强、做大，近年来共发放扶持资金 4 500 万元；养殖户增产增收，仅 2005 年一年长海地区养殖户就增收 1 亿多元，科研部门的成果得以迅速转化，银行则可以得到丰厚的利益回报。这种多赢的结局，无论在形式上还是内容上都是海洋渔业产业化发展模式的一种创新。经过多年实践与探索，被证明是一种适应海洋开发产业特点的、行之有效的创新发展模式，也是海洋渔业产业化未来发展所依赖的主要路径。

（二）"渔户＋专业合作组织＋企业"新模式

随着中国社会主义市场经济体制的不断完善，广大渔户在家庭承包经营基础上，自愿组织、自主兴办了各种专业合作组织，这是中国特色渔民合作经济组织的新形式。近年来的实践证明，渔民专业合作组织在引导渔民进入市场，发展海洋渔业产业化经营，提高水产品竞争力，促进渔民增收，建设现代海洋渔业等方面发挥了积极作用。

2005年中央1号文件提出引导渔民专业合作组织发展水产品加工业，为创新产业化经营模式带来了契机。渔民专业合作组织发展水产品加工业，既可以把水产品的加工、销售所获得的利润留在海洋渔业内部，达到海洋渔业增效、渔民增收的目的，又可以形成"渔户＋专业合作组织＋企业"的模式，实现产业化经营模式的创新。由于专业合作组织是由渔户组成的，而企业又是由专业合作组织建立的，因此所形成的"渔户＋专业合作组织＋企业"的模式的利益联接机制是比较完善的，企业与渔户的利益联接是紧密的，而不是单纯地追求各自利益的最大化。正是由于加工企业是由渔户组成的合作经济组织建立和拥有的，从而确保了企业与渔户、专业合作组织在利益上的一致性，也就较好地避免了以往产业经营模式在利益联接上存在的弊端。正因为如此，这一模式将会有很好的发展前景。

但从总体上看，目前大多数的合作组织仍存在着规模较小、实力不强、运作不规范、运行质量较差、不够稳定等问题，很多合作组织缺少自有资金，又很少有固定资产，特别是缺乏加工转化能力，因此在拓展市场、占领市场上，难以获得较好的效果。如何克服合作组织的这些弊端，突破其发展的瓶颈，将是下一步完善海洋渔业产业化模式所应该努力的方向，也才会使这一模式发挥真正的作用。

（三）"超市＋水产品加工企业＋渔户"新模式

随着人们生活水平的提高和环保意识的觉醒，广大消费者对水产品的质量安全也提出了较高的要求。水产品的质量安全程度同水产品的供应链模式有着密切的联系，而目前中国的水产品供应链存在着以下两个问题。第一，改革开放以来，中国水产品供应链上的经营主体便以小规模的渔户生产和流通部门中的个体户商贩为主，这些经营主体过于分散、规模太小的特点，导致其难以采用比较先进的经营理念和技术来提高水产品的质量安全程度。第二，即使在生产阶段确保了水产品的质量安全，但由于质量安全水产品的生产成本往往会高于普通水产品。通过什么方式才能使消费者认可质量安全水产品，并使那些愿意消费质量安全水产品的消费者付出更高的价格来购买，从而解决水产品的销路问题就成为了突出的问题。

中国超市的出现和迅速发展为解开质量安全水产品销售难题创造了客观条件。自从20世纪90年代以来中国超市获得迅速发展，目前已经逐步成为城市消费者购物的主要场所。超市具备多种经营质量安全水产品的客观条件：①拥有先进的管理理念及物流、销售设备和技术；②拥有可靠的标识系统和单品管理和跟踪系统；③外部竞争环境促使超市经营质量安全农产品。

这一模式的优势在于，满足了人们对食品质量安全要求越来越高这一新的要求，并充分意识到：一方面，消费者希望从超市购买到质优、安全的水产品；另一方面，有能力生产安全水产品的加工企业和业户也正在寻找零售市场。因此，在三者之间建立紧密的联结，把愿意生产安全水产品的渔业户和有能力经营安全水产品的加工企业及选择销售安全水产品的超市联接起来，加上政府部门的监督和媒体的配合，高质量的安全水产品的"卖"和"买"难问题就解决了。

实践中，已经有很多超市同水产品加工企业建立了紧密的合作关系，通过水产品加工企业带动分散的小规模渔户来生产质量安全的水产品。为了加快完善中国质量安全水产品供应体系，建议社会各方面在继续加强水产品产业链建设的同时，着重加强一元化安全管理，充分强化海洋渔业产业化在水产品质量安全管理过程中的积极作用，把超市纳入"海洋渔业产业化龙头企业"的范围，鼓励发展"超市+水产品加工企业（或渔民协会）+渔户"的新型海洋渔业产业化经营模式。打破部门之间的行政分割，建立跨越商务部门和海洋渔业部门的海洋渔业产业化管理机构，以便于支持和管理新型的"超市+水产品加工企业（或渔民协会）+渔户"的海洋渔业产业化经营模式。同时，有关政府部门也需要制定相关的政策鼓励和支持水产品加工企业把产品送入超市，为水产进超市创造条件。总的来说，针对提高水产品质量安全的这一模式是大势所趋，应当积极倡导。

四、小结

海洋渔业产业化是中国海洋渔业发展的有效途径，同时，海洋渔业产业化经营模式也是发展中国海洋渔业的新课题。已有的龙头企业带动型、主导产业带动型、市场带动型、中介组织带动型等发展模式在促进海洋渔业增效、渔民增产方面发挥了较大的作用，但也存在如利益联接机制不够紧密、主导产业低水平重复建设等问题。为了有效地解决这些问题，需要在实践的基础上积极探索更加成熟的产业化发展模式。目前探索比较成功的有"五合一"的海洋渔业产业化新模式、"渔户+专业合作组织+企业"新模式和"超市+水产品加工企业+渔户"新模式。这三种模式将以其更加全面的效能，更加紧密的利益联接机制和更加完善的理念而成为今后海洋渔业产业化发展的新方向，应积极地实践和推行。

参考文献

[1] 王淼，潘学峰.海洋渔业产业化的发展模式及运行机制［J］.中国渔业经济，2003，(6).

[2] 权锡鉴.我国海洋渔业产业化的推进策略［J］.农业经济，2002 (12).

[3] 杨江峰.罗红彬.农业产业化模式探论［J］.西北农林科技大学学报（社会科学版），2003 (2).

Discussion on development mode of Chinese ocean fishery industrialization

Shao Guilan, Lee Hunhyeon, Zhang Xi

College of Economy, Ocean University of China, Qingdao, 266071

【Abstract】 Fishery industrialization plays an important part in the change of fishery development strategy. Nowadays, there are four kinds of development modes in our fishery industrialization: driven by leading enterprises, driver by main industry, driven by market and driven by intermediary organization. We need to innovate development mode constantly to satisfy demand of productivity. For example, "five joint in one" mode, "fishermen + professional cooperation organization + enterprises" mode and "supermarket + aquatic products processing enterprises + fishermen" mode, they will become advanced development modes for their more mature operation way.

【Key Words】 Ocean fishery industrialization　Development mode　Innovation

JEL Classification: Q22

《中国海洋经济评论》[2008卷] 征稿启事

《中国海洋经济评论》(Ocean Economics Review of China, OERC) 是由中国海洋大学经济学院海洋经济研究中心、教育部人文社科重点研究基地中国海洋大学海洋发展研究院海洋经济研究所主办的海洋经济领域专业学术文献,由经济科学出版社每年定期出版和发行,旨在发表中国以及国际海洋经济领域的原创性研究成果(包括论文、书评、会议综述等),2008年度侧重发表包括但不局限于以下研究内容:

- 中国海洋经济研究(专题研究或综合研究)30年总结与回顾
- 海洋经济理论与方法论建设进展
- 海洋资源国际争夺与开发
- 海洋环境变化(包括海洋灾害)的经济影响评价
- 海洋生态系统服务价值评估
- 海洋资源(尤其是极地和大洋海底区域资源)竞争与海洋政策国际协调
- 区域港口群演化与博弈
- 涉海企业全球价值链治理与区域海洋产业升级
- 国家及区域海洋创新体系建设与海洋创新平台建设

《中国海洋经济评论》篇幅一般为10 000~15 000字,文章撰写规范请参照《中国工业经济》,并且文章具体要求一般涵盖以下内容:

- 论文题目、作者姓名(文章后附作者简介)、工作单位、邮政编码、电子邮箱
- 中英文摘要(200字以上)
- 中英文关键词(3~5个)
- 正文
- 参考文献(采用顺序编码著录方式)

投稿文章要求为汉语或者英语,文章不收取版面费。投稿者请将稿件电子

版通过电子邮箱直接发给刘曙光博士（Email：dawnliu9631@263.net；或 hyfz@ouc.edu.cn）。

《中国海洋经济评论》编辑部

2008 年 4 月

Ocean Economics Review of China
Call FOR PAPERS

Ocean Economics Review of China (OERC) is a newly published academical by Marine Economy Research Center, College of Economy, and Marine Economy Research Institute, KRI Institute of Marine Development, Ocean University of China. OERC is to publish original research articles and studies that describe the latest research and developments in ocean economics in China as well as the world. Based on the topics of the first volume in 2007, we welcome research papers in the areas of interest include but are not limited to:

- Marine economy research of China in recent three decades: retrospect and prospect
- Progress of theories and methodologies of marine economics
- International competition on marine resources development
- Impact of ocean environmental change on global / national economies
- Evaluation on services of marine ecosystem
- Marine resource competition and responsibilities of international marine policies
- Cross-sea political & economic co-operations
- National / regional marine innovation system and marine innovation platform construction
- Global value chain governance and upgrading of coastal industries

OERC will be published one volume, two issues per year. Manuscripts can be sent via e-mail to **Assistant Editor of OERC** (Dr. Dawn Shuguang Liu, Email: dawnliu9631@263.net; hyfz@ouc.edu.cn). In accordance with usual practice, papers previously published and are under consideration for publication elsewhere cannot be accepted, and authors must agree not to publish accepted papers elsewhere without the prior permission of **OERC**. The paper should not normally exceed 15000 words, and should confirm to the following:

- The paper should contain the title, author (s) name, affiliation (s), mailing address, email
- Abstract describing the context and scope of the paper
- Keywords
- JEL Classification
- Main text
- References

Submission of manuscript should be written in English or Chinese. All contributions will be reviewed by two independent experts. Upon acceptance of an article, the author(s) are assumed to have transferred the copyright of the article to **OERC**. **There are no page charges to individuals or institutions for contributions.**

Editorial Board of **OERC**

March, 2008